Artificial Intelligence and Ethics

Artificial Intelligence and Ethics is a general and wide-ranging survey of the benefits and ethical dilemmas of artificial intelligence (AI). The rise of AI and super-intelligent AI has created an urgent need to understand the many and varied ethical issues surrounding the technologies and applications of AI. This book lays a path towards the benefits and away from potential risks. It includes over thirty short chapters covering the widest array of topics from generative AI to superintelligence, from regulation to transparency, and from cybersecurity to risk management. Written by an award-winning Chief Information Security Officer (CISO) and experienced Technology Leader with two decades of industry experience, the book includes real-life examples and up-to-date references. The book will be of particular interest to business stakeholders, including executives, scientists, ethicists, and policymakers, considering the complexities of AI and how to navigate these.

Tarnveer Singh, award-winning CIO/CISO. Fellow of the British Computer Society. Fellow of the Chartered Institute of Information Security. Director (Security and Compliance) at Cyber Wisdom Ltd., Experienced Chief Information Security Officer (CISO) in the Insurance sector and many other sectors.

Artificial Intelligence and Ethics

A Field Guide for Stakeholders

Tarnveer Singh

https://orcid.org/0009-0005-0854-6213

CRC Press is an imprint of the
Taylor & Francis Group, an **informa** business

Designed cover image: Shutterstock Images

First edition published 2025
by CRC Press
2385 NW Executive Center Drive, Suite 320, Boca Raton FL 33431

and by CRC Press
4 Park Square, Milton Park, Abingdon, Oxon, OX14 4RN

CRC Press is an imprint of Taylor & Francis Group, LLC

ISBN: 978-1-032-82054-5 (hbk)
ISBN: 978-1-032-81526-8 (pbk)
ISBN: 978-1-003-50270-8 (ebk)

DOI: 10.1201/9781003502708

Contents

Introduction to Artificial Intelligence

Throughout history, humanity has experienced several technological revolutions that have changed the course of civilisation. From the invention of the wheel to the Industrial Revolution, each milestone has propelled us forward, enhancing our capabilities and expanding our horizons. These revolutions have been driven by human ingenuity and the relentless pursuit of progress.

As early as 100,000 BC, stone tools were used, and the wheel was invented in 4,000 BC. The sundial was developed in 800 BC, and gunpowder revolutionised the world in the 9th century AD. The printing press in 1441 changed the world forever, and the 19th century saw the emergence of the steam engine, the railway, and the lightbulb. The 20th century saw the automobile, television, nuclear weapons, spacecraft, and the Internet. In the 21st century, biotechnology, nanotechnology, fusion and fission, and M-theory have all been developed.

The advent of artificial intelligence (AI) marks yet another significant milestone in our technological journey. AI refers to the development of computer systems capable of performing tasks that typically require human intelligence. It encompasses various subfields, including machine learning, natural language processing, and robotics. The roots of AI can be traced back to the 1950s when computer scientists first started exploring the concept of creating intelligent machines.

Machine learning is a branch of AI that focusses on developing algorithms and models that enable computers to learn and make predictions or decisions without being explicitly programmed. In other words, it is about creating systems that can automatically learn from data and improve

DOI: 10.1201/9781003502708-1

their performance over time. Machine learning algorithms can analyse large amounts of data, identify patterns, and make predictions or decisions based on those patterns.

AI has become a powerful tool in various industries, revolutionising decision-making processes and offering immense opportunities for growth and innovation. By leveraging advanced algorithms and machine learning, AI enables computers to analyse data, identify patterns, make predictions, and even learn from their experiences. We must explore the vast potential of AI and the challenges it presents for the future of decision-making.

AI goes beyond mere automation. It involves setting a desired outcome and allowing computer programs to find their own path towards achieving it. This creative capacity gives AI its power and challenges our traditional assumptions about computers and our relationship with them.

AI excels in data sorting, pattern recognition, and prediction. We can witness its impact in everyday life through translation and speech recognition services, search engines that rank websites based on relevance, and email spam filters that identify junk mail. The applications of AI are expanding rapidly, enabling innovation across all sectors of the economy.

Opportunities in AI include:

Machine learning in personalised recommendations: AI-powered machine learning algorithms are driving personalised recommendations in various services like Netflix and Amazon. These algorithms analyse users' web experiences and evolve over time, continuously learning and improving the accuracy of their recommendations. This level of personalisation enhances user experiences and drives customer satisfaction.

Smart systems for traffic management: In the United Kingdom, 'smart motorways' utilise embedded sensors and neural network systems to anticipate and manage traffic flow. These AI-powered systems gather feedback on road conditions and make real-time adjustments to optimise traffic flow. By predicting congestion and effectively managing traffic, these systems improve road safety and reduce travel times.

High-frequency trading in financial markets: AI algorithms, known as 'high-frequency trading' algorithms, enable automated responses to market conditions at a much faster pace than human traders. These algorithms analyse market data and execute trades in milliseconds, capitalising on fleeting opportunities. Similarly, financial advisers employ AI algorithms to identify investment opportunities for their clients, leveraging the speed and accuracy of AI to make informed decisions.

Improving wildlife conservation efforts: Scientists and researchers are leveraging AI to enhance wildlife conservation efforts. For example, Cornell University collaborated with machine learning specialists to develop algorithms that accurately identify the calls of right whales. By automating this process, researchers can track individual whales more effectively, contributing to the protection and preservation of endangered species.

Analysing satellite images for environmental trends: AI plays a crucial role in analysing vast amounts of satellite imagery for environmental and socio-economic trends. Machine learning algorithms identify patterns of change and development by analysing millions of digital images. This capability enables researchers and policymakers to better understand environmental changes, monitor urban development, and make informed decisions regarding resource management.

While the opportunities presented by AI are vast, here is a brief introduction to some of the significant challenges that need to be addressed to ensure its successful implementation.

Ethical considerations: AI raises ethical concerns, such as the potential for biased decision-making or the ethical implications of AI-driven automation. As AI algorithms learn from existing data, they may perpetuate existing biases present in the data. It is crucial to develop ethical guidelines and frameworks that ensure fairness, transparency, and accountability in AI systems.

Data privacy and security: AI relies heavily on data, often personal and sensitive, to make accurate predictions and decisions. Ensuring data privacy and security is of paramount importance to prevent unauthorised access, data breaches, and misuse of personal information. Organisations must implement robust data protection measures and comply with relevant regulations to maintain trust and protect user privacy.

Lack of transparency in AI algorithms: Deep learning algorithms, such as deep neural networks, operate with multiple layers of statistical operations that are defined by algorithms rather than human input. This makes the decision-making process of AI systems complex and difficult to interpret. Ensuring transparency in AI algorithms is crucial for building trust and understanding how decisions are made.

Adapting workforce skills: The widespread adoption of AI will inevitably impact the workforce. While AI can automate repetitive and mundane tasks, it also requires individuals with the skills to develop, implement, and manage AI systems. Organisations must invest in upskilling and reskilling their workforce to adapt to the changing landscape and leverage the opportunities presented by AI.

As mentioned earlier, AI algorithms can inadvertently perpetuate biases present in the data they learn from. This can lead to discriminatory outcomes and reinforce existing social inequalities. It is essential to address bias in AI systems through rigorous testing, diverse training data, and continuous monitoring to ensure fairness and eliminate discriminatory practices.

Alan Turing, a brilliant mathematician and computer scientist, is widely regarded as one of the founding fathers of AI. Born in 1912, Turing's work laid the groundwork for modern computing and revolutionised the way we think about machines and their ability to imitate human intelligence. His contributions to AI were not only ground-breaking but also sparked ongoing ethical debates that continue to shape the field today.

Turing's most notable achievement was the concept of the '*Turing Machine*', a theoretical device that could simulate any algorithmic computation. This concept served as the basis for the development of modern computers and laid the foundation for the field of AI. Turing's visionary ideas pushed the boundaries of what was thought possible, leading to the creation of intelligent machines capable of learning and problem-solving.

Alan Turing recognised the potential ethical implications of AI early on. In his ground-breaking paper '*Computing Machinery and Intelligence*', Turing introduced the concept of the *Turing Test*. The Turing Test proposes that a machine can be considered intelligent if it can successfully deceive a human into believing that it is also human. This test raises questions about the nature of consciousness, the boundaries of machine intelligence, and the ethical implications of creating machines that can imitate humans.

Turing believed that machines could possess intelligence, but he also emphasised the importance of understanding the limitations of AI. He argued that even if a machine could pass the Turing Test, it would still lack the emotional and subjective experiences that humans possess. This distinction between human and machine intelligence is crucial in ethical discussions surrounding AI and helps shape our understanding of the unique ethical challenges AI presents.

The Turing Test has profound implications for AI and the ethical considerations surrounding it. If a machine can convincingly pass the Turing Test, it raises questions about the moral status and rights of intelligent machines. Should machines be granted the same rights and protections as humans? Can machines have moral agency and be held accountable for their actions?

These questions become increasingly relevant as AI technologies advance. While machines may not possess consciousness in the same way humans do, they can still have significant impacts on human lives. It is essential to consider the ethical implications of creating intelligent machines and to develop guidelines and regulations that ensure responsible and ethical use of AI.

This book will explore the benefits of AI and how business leaders can take advantage of it. It will also explore ethical considerations and how an ethical framework can help support organisations to navigate safely the advantages of AI. AI offers immense opportunities for growth and innovation across various industries. From personalised recommendations to traffic management and wildlife conservation, AI is transforming decision-making processes. However, challenges such as ethical considerations, data privacy, transparency, workforce adaptation, and bias need to be addressed to harness the full potential of AI while ensuring fairness and accountability. By navigating these challenges and embracing AI responsibly, we can unlock a future where intelligent machines enhance our lives and enable us to make more informed decisions.

https://orcid.org/0009-0005-0854-6213

CHAPTER 1

The Benefits of Artificial Intelligence (AI)

Artificial intelligence (AI) is no longer just a concept from science fiction; it has become a reality that is revolutionising various industries and delivering tangible benefits. AI refers to the development of computer systems that can perform tasks that typically require human intelligence, such as speech recognition, decision-making, learning, and problem-solving.

AI is based on the idea of creating machines that can simulate human intelligence. It involves the development of algorithms and models that enable computers to process and analyse vast amounts of data, recognise patterns, and make predictions or decisions based on the information at hand. AI can be categorised into two types: narrow AI, which is designed to perform specific tasks, and general AI, which has the ability to understand, learn, and apply knowledge across various domains.

The concept of AI has been depicted in science fiction for decades, with films like "2001: A Space Odyssey" and "The Matrix" showcasing intelligent machines that can interact with humans. However, it is only in recent years that AI has made significant progress and become a practical solution for real-world problems. Advances in computing power, big data, and machine learning algorithms have accelerated the development and adoption of AI technologies.

DOI: 10.1201/9781003502708-2

AI is transforming various industries by enabling automation, improving efficiency, and driving innovation. Therefore, we must explore some of the sectors that are benefiting from the implementation of AI:

AI-powered solutions in healthcare: In the healthcare industry, AI is revolutionising patient care, diagnosis, and treatment. AI algorithms can analyse medical records, images, and genetic data to detect patterns and make predictions about diseases. This helps doctors in making accurate diagnoses and provides personalised treatment plans. AI-powered robotic surgeons are also being used to perform complex surgeries with precision and minimal invasiveness.

AI's impact on the finance and banking sector: AI is transforming the finance and banking sector by automating processes, improving risk assessment, and enhancing customer experience. AI-powered chatbots and virtual assistants are being used to provide personalised financial advice and support. Machine learning algorithms are also utilised for fraud detection, credit scoring, and algorithmic trading, which improves efficiency and reduces the risk of human error.

AI in manufacturing and automation: In the manufacturing industry, AI is driving automation and improving production efficiency. AI-powered robots and machines can perform repetitive tasks with precision and speed, leading to higher productivity and reduced costs. AI algorithms can also optimise supply chain management, predict equipment failures, and improve quality control processes.

AI revolutionising customer service and support: AI is transforming customer service and support by providing instant and personalised assistance. Chatbots and virtual assistants can handle customer enquiries, provide product recommendations, and resolve issues, improving customer satisfaction and reducing wait times. Natural language processing algorithms enable these AI systems to understand and respond to customer queries in a human-like manner.

AI in the transportation and logistics industry: AI is revolutionising the transportation and logistics industry by enabling autonomous vehicles, optimising route planning, and improving logistics operations. Self-driving cars and trucks powered by AI algorithms have the potential to increase road safety and reduce traffic congestion. AI can also analyse data from sensors and Inter of Things (IoT) devices to optimise the transportation of goods, minimising costs and improving delivery times.

AI's role in the education sector: In the education sector, AI is transforming the way students learn and teachers teach. AI-powered educational tools can personalise learning experiences, adapt to students' individual needs, and provide real-time feedback. AI algorithms can also analyse student data to identify learning gaps and recommend targeted interventions.

Additionally, AI-powered virtual reality and augmented reality technologies are being used to create immersive and interactive learning environments.

The implementation of AI in various industries brings numerous tangible benefits:

Increased efficiency and productivity: AI automates repetitive tasks, allowing employees to focus on higher-value activities. This leads to improved efficiency and productivity in organisations.

Enhanced decision-making: AI algorithms can analyse vast amounts of data and provide insights that aid in decision-making. This enables organisations to make informed and data-driven choices, leading to better outcomes.

Improved accuracy and precision: AI systems can perform tasks with a high degree of accuracy and precision, reducing errors and improving the quality of outputs. This is particularly beneficial in industries such as healthcare and manufacturing, where precision is crucial.

Cost savings: AI-powered automation can reduce labour costs and operational expenses. By streamlining processes and optimising resource allocation, organisations can achieve significant cost savings.

Enhanced customer experience: AI enables personalised and responsive customer interactions, leading to improved satisfaction and loyalty. AI-powered chatbots and virtual assistants can provide instant and round-the-clock support, enhancing the customer experience.

AI offers significant benefits, but there are significant challenges and ethical considerations that need to be addressed:

Job displacement: The automation of tasks through AI technologies may result in job displacement, particularly for roles that can be easily automated. It is important to consider the impact on the workforce and work towards reskilling and upskilling employees to adapt to the changing job landscape.

Data privacy and security: AI relies on vast amounts of data, raising concerns about data privacy and security. Organisations must ensure that appropriate measures are in place to protect sensitive information and comply with data protection regulations.

Bias and fairness: AI algorithms can be biased if trained on biased data or if biased assumptions are built into the models. It is crucial to address bias and ensure fairness in AI systems to avoid perpetuating existing societal inequalities.

Ethical decision-making: AI systems may need to make ethical decisions, such as in autonomous vehicles where decisions about potential accidents need to be made. There is a need for ethical frameworks and guidelines to ensure that AI systems make decisions that align with societal values.

The future of AI holds immense potential for further revolution across industries. As AI technologies continue to advance, we can expect:

Continued automation: AI will drive further automation across industries, leading to increased efficiency and reduced costs. This will free up human resources for more creative and complex tasks.

Improved personalisation: AI will enable even more personalised experiences in various domains, such as healthcare, education, and entertainment. AI systems will understand individual preferences and adapt to provide tailored recommendations and solutions.

Enhanced collaboration between humans and machines: AI will enable humans and machines to collaborate more effectively, combining their respective strengths. This will lead to new opportunities for innovation and problem-solving.

Ethical AI development: As AI becomes more prevalent, there will be an increased focus on ethical AI development. Organisations and policymakers will work towards creating frameworks and regulations that ensure the responsible and ethical use of AI technologies.

AI has moved from the realms of science fiction to become a reality that is revolutionising industries and delivering tangible benefits. From healthcare to finance, manufacturing to customer service, AI is transforming the way we work and live. While there are challenges and ethical considerations to address, the future of AI holds immense potential for further revolution and positive impact. It is essential that we embrace AI responsibly and harness its power to create a better and more sustainable future.

https://orcid.org/0009-0005-0854-6213

CHAPTER 2

Machine Learning and Automation

MACHINE LEARNING IS AN area of artificial intelligence (AI) that focusses on developing algorithms and models that allow computers to learn from and make predictions or decisions based on data. Machine learning is the driving force behind the rapid advancements in AI. It involves training algorithms to recognise patterns and make decisions or predictions without explicit programming. By feeding large amounts of data into these algorithms, machines can learn from experience and improve their performance over time. There are three main types of machine learning: supervised learning, unsupervised learning, and reinforcement learning.

In supervised learning, algorithms are trained using labelled data, where the desired output is known. This enables the algorithm to learn from the examples and make predictions on unseen data. Unsupervised learning, on the other hand, deals with unlabelled data, where the algorithm needs to find patterns or structure in the data without any predefined labels.

Reinforcement learning involves training an algorithm to interact with an environment and learn by trial and error, receiving rewards or penalties based on its actions.

DOI: 10.1201/9781003502708-3

Machine learning has found applications in a wide range of industries, revolutionising the way businesses operate and making significant contributions to fields such as healthcare, finance, transportation, and entertainment. In healthcare, machine learning algorithms are being used to diagnose diseases, predict patient outcomes, and personalise treatments based on individual characteristics. In finance, machine learning is employed for fraud detection, algorithmic trading, and risk assessment. Machine learning algorithms also play a vital role in improving transportation systems, optimising logistics, and enhancing customer experiences in the entertainment industry.

The ability of machine learning algorithms to process and analyse vast amounts of data quickly and accurately has opened up new possibilities for businesses in various sectors. By harnessing the power of machine learning, organisations can gain valuable insights, make data-driven decisions, and automate processes, leading to improved efficiency, reduced costs, and enhanced customer satisfaction.

As machine learning continues to advance, its impact on society is becoming increasingly profound. On the positive side, machine learning has the potential to improve the quality of life, enhance productivity, and drive economic growth. It can revolutionise healthcare by enabling early disease detection and personalised treatments, optimise transportation systems to reduce congestion and emissions, and facilitate scientific breakthroughs by analysing complex datasets.

However, machine learning also raises concerns and ethical considerations. The use of AI-powered algorithms in decision-making processes, such as hiring, loan approvals, or criminal justice, can introduce bias and perpetuate discriminatory practices if not carefully monitored. Privacy and data security are also significant concerns, as the collection and analysis of vast amounts of personal data become more prevalent. It is crucial to ensure that machine learning is transparent, explainable, and accountable and that appropriate regulations and safeguards are in place to mitigate potential risks.

While the potential of machine learning is immense, there are several challenges that need to be addressed. One of the challenges is the availability of high-quality and unbiased training data. Machine learning algorithms heavily depend on the quality and diversity of the data they are trained on. Biased or incomplete datasets can lead to biased predictions or reinforce existing societal prejudices. Another challenge is the interpretability and explainability of machine learning models. Many machine learning algorithms, such as deep neural networks, are often referred to as "black boxes" because they produce highly accurate results but lack transparency. This makes it difficult to understand how decisions are made or to detect and correct potential errors or biases.

The future of machine learning and AI is incredibly promising. One area of significant development is the exploration of deep learning techniques, inspired by the structure and function of the human brain. Deep learning has shown remarkable success in areas such as image recognition, natural language processing, and speech recognition. Continued research in this field is likely to yield breakthroughs in complex problem-solving and decision-making.

In recent years, machine learning algorithms have made significant advancements, enabling machines to perform complex tasks with remarkable accuracy. One such breakthrough is the rise of generative adversarial networks (GANs), a class of algorithms that can generate realistic and original content, such as images, music, or text. GANs have the potential to revolutionise creative industries and pave the way for new forms of artistic expression.

Another noteworthy development is the integration of machine learning with other emerging technologies, such as robotics and the Internet of Things (IoT). By combining machine learning with robotics, we can create intelligent autonomous systems capable of performing tasks in dynamic and unpredictable environments. In the realm of IoT, machine learning algorithms can analyse vast amounts of sensor data to optimise energy consumption, detect anomalies, and improve overall efficiency.

Several industries have embraced machine learning and are leading the way in its adoption. The healthcare industry, for example, is leveraging machine learning algorithms to improve patient outcomes, streamline clinical workflows, and accelerate drug discovery. Financial institutions are using machine learning to detect fraud, automate risk assessment, and personalise financial services. E-commerce companies are employing machine learning algorithms to enhance customer experiences, recommend products, and optimise pricing strategies.

Machine learning is also making significant strides in the automotive industry, where it is enabling the development of self-driving cars and enhancing vehicle safety. Additionally, the entertainment industry is leveraging machine learning to personalise content recommendations, analyse audience preferences, and create immersive virtual reality experiences.

Machine learning has become a ubiquitous part of our everyday lives, often without us even realising it. From the personalised recommendations on streaming platforms to voice assistants like Siri and Alexa, machine learning algorithms are continuously working behind the scenes to make our lives more convenient and enjoyable.

In the realm of healthcare, wearable devices equipped with machine learning algorithms can monitor vital signs, detect irregularities, and provide timely alerts. Smart home systems use machine learning to learn our preferences and automate various tasks, such as adjusting the thermostat or turning on the lights. Machine learning algorithms also power virtual assistants, enabling us to interact with our devices through natural language processing.

Automation is fast becoming an important means to streamline processes, increase efficiency, and drive innovation. Automation offers a myriad of benefits to industries seeking to enhance their efficiency. One of the key advantages is the elimination of manual and repetitive tasks. By automating mundane processes, businesses can allocate their human resources to more strategic and impactful activities. This not only saves time but also reduces the risk of human error, resulting in improved accuracy and productivity.

Automation enables companies to operate around the clock without the need for continuous human intervention. This 24/7 functionality ensures uninterrupted operations, accelerates turnaround times, and enhances customer satisfaction. Additionally, automation facilitates real-time data collection and analysis, allowing businesses to make informed decisions and respond promptly to market demands.

The adoption of automation is rapidly increasing in industries worldwide. This trend is driven by the desire to reduce costs, enhance productivity, and gain a competitive edge. In terms of specific automation technologies, robotic process automation (RPA) is gaining significant traction. RPA involves the use of software robots that mimic human actions to perform repetitive tasks. By leveraging RPA, companies can automate processes across departments, such as finance, human resources, and customer service. This not only reduces manual errors but also frees up employees to focus on more strategic activities.

Automation encompasses a wide range of technologies that cater to different industry needs. One such technology is machine learning, a subset of AI, which enables systems to learn from data and improve their performance over time. Machine learning algorithms can analyse vast amounts of information to identify patterns and make predictions, aiding in tasks such as fraud detection, demand forecasting, and personalised recommendations.

Another prominent automation technology is robotic automation, which involves the use of physical robots to perform tasks traditionally done by humans. These robots are equipped with sensors and advanced algorithms that enable them to operate autonomously, navigate complex environments, and interact with objects. Industries like manufacturing, logistics, and healthcare have witnessed significant advancements through the deployment of robotic automation.

The implementation of automation varies across industries, with each sector leveraging automation technologies to address unique challenges and opportunities. In manufacturing, for instance, automation has revolutionised production lines by reducing cycle times, enhancing quality control, and enabling mass customisation. Automated robots efficiently perform repetitive tasks, while machine learning algorithms optimise supply chain management and predictive maintenance.

In the healthcare industry, automation has paved the way for improved patient care and streamlined administrative processes. Robotic surgery systems enable precise and minimally invasive procedures, reducing the risk of complications and speeding up recovery times.

Automated appointment scheduling, medical record management, and billing systems streamline administrative tasks, freeing up healthcare professionals to focus on patient care.

Numerous case studies highlight the successful implementation of automation across industries. One such example is Amazon's use of automation in its fulfilment centres. The company employs a vast network of robots that work alongside human employees to pick, pack, and ship products. This automation has significantly expedited order fulfilment, improved inventory management, and reduced operational costs, allowing Amazon to meet the demands of its rapidly growing customer base.

Another noteworthy case study is Tesla's implementation of automation in its car manufacturing process. The company utilises an extensive array of robotic systems to assemble vehicles, reducing production time and improving quality control. Tesla's innovative approach to automation has not only revolutionised the automotive industry but has also set new standards for efficiency and sustainability.

While the benefits of automation are undeniable, its adoption does not come without challenges. One of the primary concerns is the potential displacement of human workers. As automation takes over repetitive tasks, businesses must focus on retraining and upskilling their workforce to take on more complex and creative roles. This transition requires effective change management strategies and a commitment to lifelong learning.

Another consideration is the integration of automation technologies with existing systems and processes. Seamless integration is crucial to ensure a smooth transition and avoid disruptions in operations. Additionally, industries must address cybersecurity concerns, as automation opens up new avenues for potential threats and vulnerabilities. Robust security measures, including encryption, access controls, and regular audits, are essential to safeguard sensitive data and systems.

AI and machine learning play a pivotal role in automation. These technologies enable systems to learn, adapt, and make intelligent decisions based on vast amounts of data. AI-powered automation not only enhances efficiency but also enables predictive and prescriptive analytics, leading to more accurate forecasting and decision-making.

Machine learning algorithms are particularly effective in automating complex tasks that require pattern recognition and data analysis. For example, in the finance industry, AI-powered algorithms can automate fraud detection by analysing transaction patterns and identifying anomalies in real-time. Similarly, in customer service, chatbots powered by AI can provide instant and personalised assistance, improving customer satisfaction and reducing the need for human intervention.

One of the emerging trends is the convergence of automation with other transformative technologies such as the IoT and big data analytics. This integration enables the automation of entire ecosystems, where interconnected devices and systems collaborate seamlessly to optimise operations and enhance user experiences.

Additionally, the rise of edge computing is set to revolutionise automation by bringing processing power closer to the source of data generation. This decentralised approach reduces latency, improves real-time decision-making, and enhances security. Edge computing combined with automation opens up new avenues for industries such as autonomous vehicles, smart cities, and precision agriculture.

Automation has become an integral part of industries worldwide, driving efficiency, innovation, and growth. From RPA to machine learning and AI-powered systems, automation technologies continue to evolve and transform various sectors. As industries embrace automation, they must navigate challenges such as workforce reskilling, system integration, and cybersecurity.

Looking ahead, automation will continue to shape the future of industries. Advancements in AI, machine learning, and other emerging technologies will enable more sophisticated automation, leading to increased productivity, enhanced decision-making, and improved customer experiences. As businesses adapt to this new era of automation, those that embrace and harness its potential will thrive in the rapidly evolving digital landscape.

https://orcid.org/0009-0005-0854-6213

CHAPTER 3

Implementing Generative AI in Your Business

ARTIFICIAL INTELLIGENCE (AI) has revolutionised numerous industries, and its potential for organisations is boundless. Generative AI (GenAI), a cutting-edge technology that combines genetics and AI, offers a myriad of hidden benefits that can propel your organisation to new heights. Directors must explore the untapped advantages of GenAI and how it can transform your organisation.

GenAI harnesses the power of AI to analyse and interpret vast amounts of genetic data. By leveraging machine learning algorithms, it can provide valuable insights into genetic traits and health conditions and even predict future outcomes. This level of understanding has never been possible before, and it opens up a world of possibilities for organisations across various sectors.

The hidden benefits of GenAI lie in its ability to uncover hidden patterns in genetic data, enabling organisations to make more informed decisions. By analysing genetic information, organisations can identify potential health risks among their workforce, enabling them to take proactive measures to mitigate these risks. Additionally, GenAI can help organisations identify genetic traits that may be advantageous for certain roles, such as enhanced memory or problem-solving skills. By leveraging

DOI: 10.1201/9781003502708-4

these genetic insights, organisations can optimise their workforce and maximise productivity.

One of the key benefits of GenAI for organisations is its ability to boost productivity and efficiency. By analysing genetic data, organisations can gain valuable insights into the unique strengths and weaknesses of their employees. This information can be used to optimise work assignments, ensuring that employees are placed in roles that align with their genetic predispositions. This alignment leads to higher job satisfaction, increased motivation, and, ultimately, enhanced productivity.

Moreover, GenAI can identify genetic factors that affect energy levels and sleep patterns. By understanding these factors, organisations can implement strategies to optimise employee schedules and create a healthier work environment. For example, employees who are genetically predisposed to be more productive in the morning can be assigned tasks that require higher concentration during that time. This level of customisation not only boosts productivity but also improves work-life balance for employees.

In today's fast-paced business environment, making informed decisions is crucial for success. GenAI can significantly enhance decision-making processes within organisations. By analysing genetic data, organisations can gain insights into employees' cognitive abilities, decision-making styles, and risk tolerance levels. This information can be invaluable when making important strategic decisions.

For instance, GenAI can identify individuals who possess a higher risk tolerance and are more likely to make bold decisions. These individuals can be assigned to roles that require innovative thinking and a willingness to take risks. On the other hand, employees who are more risk-averse can be assigned to roles that require careful analysis and decision-making.

By leveraging these genetic insights, organisations can create a diverse and balanced decision-making team, leading to more effective and successful outcomes.

Customer experience is a critical factor in today's competitive business landscape. GenAI can play a pivotal role in improving customer experience by personalising products and services based on genetic information. By analysing genetic data, organisations can gain insights into customers' preferences and health conditions and even predict their future needs.

For example, a healthcare organisation can use GenAI to offer personalised treatment plans based on an individual's genetic predispositions. This level of customisation not only improves the effectiveness of treatment but also enhances the overall patient experience.

Similarly, retailers can use GenAI to recommend products that align with customers' genetic traits, leading to higher customer satisfaction and loyalty.

In addition to its ability to improve productivity and decision-making, GenAI can streamline processes and reduce costs for organisations. By analysing genetic data, organisations can identify genetic traits that are associated with specific skills or talents. This information can be used to optimise recruitment and talent acquisition processes, ensuring that the right candidates are selected for the right roles.

Furthermore, GenAI can identify genetic factors that affect employee performance and well-being. By understanding these factors, organisations can implement targeted wellness programmes and interventions, leading to reduced absenteeism, improved employee satisfaction, and lower healthcare costs. By leveraging these hidden benefits of GenAI, organisations can optimise their processes, reduce costs, and gain a competitive edge in the market.

Implementing GenAI in your organisation may come with its fair share of challenges. One of the key challenges is ensuring data privacy and security. Genetic information is highly sensitive, and organisations must invest in robust security measures to protect this data.

Additionally, organisations may face resistance from employees who are concerned about the ethical implications of genetic analysis. It is crucial to address these concerns through transparent communication and comprehensive employee education programmes.

To successfully implement GenAI, organisations must invest in training and upskilling their employees. This technology requires a deep understanding of genetics and AI, and organisations should provide the necessary resources and training programmes to ensure employee competency. By investing in the development of their workforce, organisations can overcome challenges and fully harness the benefits of GenAI.

To ensure a smooth transition to GenAI adoption, organisations must prioritise training and upskilling their employees. GenAI requires a multidisciplinary skill set, combining genetics, AI, and data analysis. Organisations should provide comprehensive training programmes that equip employees with the necessary knowledge and skills to leverage GenAI effectively.

Additionally, organisations should foster a culture of continuous learning and encourage employees to stay updated with the latest advancements in GenAI. By investing in the development of their workforce, organisations can create a knowledgeable and empowered team that can fully harness the potential of GenAI.

The field of GenAI is rapidly evolving, and there are exciting future trends and advancements on the horizon. One such trend is the integration of GenAI with wearable technology. By combining genetic data with real-time health and activity data from wearables, organisations can gain deeper insights into employee well-being and performance. This integration has the potential to revolutionise workplace wellness programmes and optimise employee engagement and productivity.

Another future trend is the use of GenAI in personalised medicine. As our understanding of genetics improves, GenAI can be used to develop targeted treatment plans and medications based on an individual's genetic makeup. This personalised approach to medicine has the potential to transform healthcare and improve patient outcomes significantly.

GenAI offers a wealth of hidden benefits for organisations, from boosting productivity and efficiency to enhancing decision-making and improving customer experience. By leveraging the power of AI and genetics, organisations can unlock a new level of understanding and optimise their processes and workforce.

While implementing GenAI may come with its challenges, organisations can overcome them through robust data security measures and comprehensive employee training programmes. By investing in the development of their workforce and staying updated with future trends and advancements, organisations can harness the full potential of GenAI and gain a competitive edge in the ever-evolving business landscape.

Generative AI, also known as GenAI, has gained significant attention in recent years. Although many organisations are experimenting with AI techniques, only a small percentage have successfully implemented them across multiple business units and processes. Any business considering adopting GenAI can explore the four key steps to implementing generative AI in your enterprise.[1]

Establish your vision for GenAI: To effectively implement GenAI, it is crucial to align your objectives with your enterprise goals. Begin by restating your corporate vision and clearly articulating how AI will support that vision. For example, AI can enable better business value by improving customer satisfaction, reducing costs, and driving top-line revenue growth. By linking GenAI objectives to your enterprise goals, you can ensure that your AI initiatives are aligned with your overarching vision.

Prioritise adoption: Not all GenAI initiatives are equal in terms of value and feasibility. It is essential to prioritise adoption based on the potential benefits and the feasibility of each initiative. Engage both IT and business leaders in the decision-making process to identify the most valuable and feasible initiatives. By prioritising adoption, you can focus your resources on initiatives that offer the greatest value and have a higher likelihood of success.

Remove barriers to capturing value: Implementing GenAI can be hindered by various organisational barriers. Identify these barriers and take the necessary actions to overcome them. For example, aligning projects with corporate goals increases the chances of success and maturity. Establish formal structures of accountability to ensure the credibility of project maturity. By addressing these barriers, you can create an environment that is conducive to capturing the full value of GenAI.

Identify the risks: Like any technology, GenAI comes with its own set of risks. It is crucial to identify and assess these risks to develop effective mitigation strategies. Some of the risks associated with GenAI include regulatory compliance, reputational damage, competency gaps, and technical debt. Collaborate with cross-functional teams, including legal, risk, and security experts, to evaluate the feasibility of use cases and mitigate the identified risks.

To facilitate the successful implementation of GenAI, it is essential to consider the key components of your AI strategy framework. These components include:

- Vision
    - *Goals*: Clearly define the goals you aim to achieve through GenAI implementation.
    - *Benefits*: Identify the potential benefits that GenAI can bring to your enterprise.
    - *Success metrics*: Establish metrics that align with your business goals to measure the success of individual use cases.
- Value
    - *Business impact*: Evaluate the potential impact of GenAI on your business processes and outcomes.

- *Change management*: Develop strategies to manage the organisational changes that GenAI implementation may bring.
- *People and skills*: Assess the skills and capabilities required to effectively utilise GenAI within your organisation.

- Risks
  - *Regulatory*: Stay updated on evolving regulatory landscapes and ensure compliance with relevant regulations.
  - *Reputational*: Take measures to protect your organisation's reputation when implementing GenAI.
  - *Competency*: Address competency gaps and invest in training and development to build AI expertise within your organisation.
- Adoption
  - *Use cases and value maps*: Identify specific use cases that align with your enterprise goals and provide tangible business value.
  - *AI decision framework*: Develop a framework to guide decision-making regarding AI initiatives.
  - *Decision governance*: Establish governance structures to ensure that AI initiatives align with your overall business strategy.

Implementing GenAI in your enterprise requires a strategic approach that is aligned with your vision, prioritises adoption, removes barriers, and mitigates risks. By following these steps and considering the key components of your AI strategy framework, you can unlock the full potential of GenAI and drive significant business value. Keep in mind that GenAI is a rapidly evolving field, and continuous learning and adaptation are essential to stay ahead in this ever-changing landscape.

NOTE

1 https://www.gartner.com/en/information-technology/topics/ai-strategy-for-business.

https://orcid.org/0009-0005-0854-6213

CHAPTER 4

Technology and Ethics

In today's rapidly evolving world, technology has become an integral part of our daily lives. From smartphones to artificial intelligence (AI), advancements in technology have revolutionised the way we live, work, and interact. However, as technology continues to advance, it brings forth a myriad of ethical implications that need to be carefully examined and addressed.

Technologists must delve into the ethical considerations surrounding the advancements in technology and explore the impact they have on various aspects of our society.

Ethics play a crucial role in guiding the development and use of technology. As technology becomes more sophisticated, it raises questions about the ethical responsibility of those involved in creating and implementing it. In the field of data privacy and security, technology has enabled the collection and analysis of vast amounts of personal information. This raises concerns about how this data is being used, who has access to it, and how it can be protected from unauthorised use. Ethical considerations become paramount in ensuring that individuals' privacy and rights are respected.

Data privacy and security have emerged as major ethical concerns in the digital age. With the proliferation of technology and the internet, personal information has become incredibly vulnerable to breaches and misuse. Companies are collecting and storing massive amounts of data, often without the explicit consent of individuals. This raises questions about the ethical responsibility of these companies to protect the privacy and security of the data they possess. Additionally, the use of data for targeted

DOI: 10.1201/9781003502708-5

advertising and surveillance purposes further exacerbates these concerns. Striking a balance between leveraging data for innovation and safeguarding individuals' privacy is a critical ethical challenge that needs to be addressed.

Advancing technology has also had a profound impact on the job market, raising ethical concerns about employment and job displacement. Automation and AI have the potential to replace a significant number of jobs, leading to unemployment and economic inequality. This raises questions about the ethical responsibility of companies and governments to ensure that technological advancements do not result in widespread job loss without providing viable alternatives. Additionally, the ethical implications of retraining and upskilling the workforce to adapt to the changing technological landscape need to be carefully considered.

AI and automation have seen remarkable advancements in recent years. While these technologies hold great promise in improving efficiency and productivity, they also raise ethical concerns. The ethical implications of AI and automation include issues such as algorithmic bias, lack of transparency, and the potential for decision-making without human intervention. It is crucial to address these concerns to ensure that AI and automation are developed and deployed in a manner that aligns with ethical principles and respects human values.

Technology has undoubtedly brought about unprecedented convenience and efficiency in various aspects of our lives. However, this convenience often comes at the expense of ethical considerations. For example, the proliferation of social media platforms has raised concerns about the impact on mental health, privacy violations, and the spread of misinformation.

Balancing convenience and ethical responsibility requires careful examination of the potential harms and benefits of technology, as well as the implementation of safeguards to mitigate any adverse effects.

Social media and online platforms have become integral parts of our daily lives, but they also present ethical challenges. The spread of fake news, cyberbullying, and the manipulation of public opinion are some of the ethical concerns associated with these platforms. Additionally, the collection and use of personal data by social media companies raise privacy and consent issues. It becomes crucial to hold these platforms accountable and ensure that ethical guidelines are in place to protect users and promote responsible behaviour.

Legislation and regulation play a vital role in addressing the ethical implications of advancing technology. Governments and regulatory bodies need to establish clear guidelines and frameworks to ensure that technology is developed and used in an ethical manner. This includes regulations surrounding data privacy, security, and the responsible use of AI. Additionally, collaboration between governments, technology companies, and other stakeholders is essential to create a comprehensive and effective regulatory framework.

Tech companies have a significant ethical responsibility in the development and deployment of technology. They must prioritise ethical considerations in their decision-making processes and ensure that their products and services do not harm individuals or society as a whole. This includes being transparent about data collection and usage, addressing algorithmic bias, and actively working to mitigate the negative impacts of technology on society. Ethical responsibility should be ingrained in the core values and practices of tech companies to foster trust and accountability.

As technology continues to advance at an unprecedented pace, it becomes increasingly important to navigate the ethical implications that arise. From data privacy and security to employment concerns and the responsible use of AI, technology ethics encompass a broad range of considerations. It is crucial for individuals, governments, and tech companies to collaborate and address these ethical challenges to ensure that technology is developed and used in a manner that benefits society as a whole. By prioritising ethics in technological advancements, we can shape a future that is both innovative and ethically responsible.

https://orcid.org/0009-0005-0854-6213

CHAPTER 5

The Technology Lifecycle

AS THE PACE OF technological advancement continues to accelerate, it is crucial for individuals and businesses alike to understand the concept of the technology lifecycle. The technology lifecycle refers to the stages that a particular technology goes through from its inception to its eventual obsolescence. By understanding these stages, we can navigate the ever-changing technology landscape more effectively and make informed decisions about when and how to adopt new technologies.

In today's fast-paced world, technology plays a central role in almost every aspect of our lives. From the way we communicate to the way we work; technology has become an integral part of our society. However, not all technologies are created equal, and not all technologies are meant to last forever. Understanding the technology lifecycle is essential because it allows us to differentiate between short-lived fads and long-lasting innovations.

By understanding the technology lifecycle, individuals and businesses can avoid investing time and resources into technologies that are already on the decline. Instead, they can focus on technologies that are in the growth or maturity stages, where there is still significant potential for innovation and adoption. Additionally, understanding the technology lifecycle helps us anticipate future trends and prepare for the next wave of technological advancements.

DOI: 10.1201/9781003502708-6

The technology lifecycle consists of several distinct stages, each with its own characteristics and challenges. The first stage is the *emergence* stage, where a new technology is introduced to the market. During this stage, the technology is often expensive, unreliable, and not widely adopted. However, it holds the promise of significant benefits and has the potential to disrupt existing industries.

As the technology matures, it enters the *growth* stage. During this stage, the technology becomes more affordable, reliable, and widely available. Adoption rates increase, and the technology starts to gain mainstream acceptance. This is the stage where early adopters and innovators play a crucial role in driving adoption and shaping the technology's future trajectory.

The next stage is the *maturity* stage, where the technology reaches its peak adoption and market saturation. In this stage, the technology is considered mainstream and has become an integral part of our daily lives. The focus shifts from innovation to optimisation and efficiency. Competition among vendors intensifies, and the market becomes saturated with similar products and services.

After the maturity stage, the technology enters the *decline* stage. This stage is characterised by decreasing adoption rates and the emergence of newer, more advanced technologies. The technology becomes outdated and is gradually replaced by newer alternatives. It is important to note that the decline stage does not mean the technology becomes obsolete overnight.

Instead, it undergoes a gradual decline until it eventually becomes obsolete.

One of the key factors that influence the technology lifecycle is the rate of technology adoption and diffusion. Technology adoption refers to the process by which individuals or organisations accept and integrate a new technology into their daily lives or operations.

Diffusion, on the other hand, refers to the spread of the technology throughout a society or market.

The rate of technology adoption and diffusion can vary significantly depending on various factors such as the complexity of the technology, its perceived benefits, and the readiness of individuals or organisations to embrace change. Innovators and early adopters are typically the first to embrace new technologies, followed by the early majority, late majority, and laggards.

Understanding the dynamics of technology adoption and diffusion is crucial for businesses looking to introduce new products or services. By identifying the target market and understanding its readiness for change, businesses can develop strategies to accelerate adoption and maximise the potential of their technologies.

The technology lifecycle is influenced by a wide range of factors, both internal and external. Internal factors include the technology's capabilities, reliability, and cost-effectiveness.

External factors include market demand, competition, regulatory environment, and societal factors.

For a technology to thrive and progress through the lifecycle, it needs to meet the needs and expectations of its users. It also needs to adapt and evolve in response to changing market dynamics and emerging trends. Technologies that fail to meet these requirements are likely to stagnate and eventually decline.

Another critical factor is the availability of resources and support for the technology. This includes a skilled workforce, research and development funding, and a supportive ecosystem of suppliers, partners, and customers. Without these resources, even the most promising technologies may struggle to gain traction and reach their full potential.

Ethics play a vital role in the technology lifecycle, particularly in the areas of privacy, security, and social impact. As technologies become more advanced and pervasive, they raise complex ethical questions that need to be addressed.

For instance, the collection and use of personal data by technology companies have raised concerns about privacy and data protection. The development of artificial intelligence and autonomous systems has raised questions about accountability and the potential for bias and discrimination. It is essential to consider these ethical implications throughout the technology lifecycle to ensure that technologies are developed and deployed in a responsible and ethical manner.

To illustrate the concept of the technology lifecycle, we can examine a few case studies of technologies in different stages.

Emergence: One example of a technology in the emergence stage is virtual reality (VR). VR has been around for decades but has recently gained significant attention and investment.

While it is still not widely adopted, VR holds immense potential in various industries, including gaming, entertainment, and education.

Growth: The smartphone is an example of a technology in the growth stage. Smartphones have become an integral part of our daily lives, with billions of people around the world using them for communication, entertainment, and productivity. The growth of the smartphone market has been driven by continuous innovation and improvements in technology.

Maturity: Personal computers are an example of a technology in the maturity stage. Personal computers have become ubiquitous, with most households and businesses owning at least one. While there is still room for innovation and improvement, personal computers have reached a point of market saturation and are considered a mature technology.

Decline: Fax machines are an example of a technology in the decline stage. With the rise of email and digital communication, fax machines have become outdated and are rarely used today. While they may still have some niche applications, their overall adoption and relevance have declined significantly.

Navigating the technology lifecycle requires careful planning and strategic decision-making. Here are some strategies to consider:

Stay informed: Keep up to date with the latest technological advancements and trends. Read industry publications, attend conferences, and engage with experts to stay informed about emerging technologies and their potential impact.

Assess relevance: Continuously evaluate the relevance of technologies to your specific needs and goals. Consider the potential benefits and risks, as well as the cost and effort required for adoption. Prioritise technologies that align with your strategic objectives and have the potential to deliver significant value.

Invest in research and development: Allocate resources to research and development to stay at the forefront of technological innovation. By investing in R&D, you can develop new technologies or enhance existing ones, ensuring their competitiveness and longevity.

Embrace collaboration: Collaborate with partners, suppliers, and customers to leverage their expertise and resources. Building strong relationships and partnerships can help accelerate adoption, drive innovation, and overcome challenges associated with the technology lifecycle.

Plan for obsolescence: Anticipate the eventual obsolescence of technologies and plan accordingly. Develop strategies for transitioning to newer technologies and ensure that your systems and processes are adaptable and flexible.

The future of technology not only holds immense promise and potential but also raises complex challenges and ethical considerations. As technologies continue to advance, they will reshape industries, transform the way we work and live, and have profound implications for society as a whole.

Artificial intelligence, blockchain, Internet of Things, and renewable energy are just a few examples of technologies that are expected to have a significant impact in the coming years. However, their success and societal acceptance will depend on how well we navigate the technology lifecycle and address the ethical implications associated with their adoption.

https://orcid.org/0009-0005-0854-6213

CHAPTER 6

Innovation and Product Management

INNOVATION is the key driver of progress in our modern society. It is the catalyst that propels us forward, enabling us to overcome challenges, improve our lives, and shape the world we live in. From the invention of the wheel to the development of artificial intelligence, innovation has been at the core of human advancement.

Throughout history, societies that have embraced liberty have witnessed remarkable advancements. From the Renaissance period in Europe, where artistic and scientific breakthroughs flourished, to the Silicon Valley of today, where pioneers in technology push the boundaries of what is possible, liberty has been a driving force behind innovation. By championing individual rights and fostering a culture of free expression, we can unlock the full potential of human ingenuity.

Innovators throughout history have exemplified the power of optimism. Thomas Edison, despite experiencing numerous failures, remained optimistic and eventually invented the electric light bulb. Elon Musk, the visionary entrepreneur behind SpaceX and Tesla, embodies the relentless optimism that drives innovation in the 21st century. By cultivating optimism within ourselves and our communities, we can inspire a culture of innovation that propels us towards a brighter future.

In a rapidly changing world, innovation is not just desirable; it is inevitable. As technology advances at an exponential pace, our lives become increasingly intertwined with new inventions and discoveries. From the

DOI: 10.1201/9781003502708-7

advent of the internet to the rise of artificial intelligence, innovation has become a constant force shaping our daily lives.

The inevitability of innovation brings both opportunities and challenges. On the one hand, innovative technologies have the potential to revolutionise industries, improve efficiency, and enhance our quality of life. On the other hand, they raise ethical concerns and require careful consideration to ensure responsible use. As we navigate this rapidly evolving landscape, it is crucial to strike a balance between embracing innovation and addressing its ethical implications.

Throughout history, numerous innovative technologies have transformed industries and revolutionised the way we live. One such example is the invention of the printing press by Johannes Gutenberg in the 15th century. This innovation democratised access to knowledge, fuelling the spread of ideas, and paving the way for the Renaissance.

In more recent times, the advent of the internet has had a profound impact on society. It has transformed the way we communicate, access information, and do business. The rise of social media platforms like Facebook and Twitter has revolutionised the way we connect with others, while e-commerce giants like Amazon have disrupted traditional retail models.

Another example is the development of renewable energy technologies. Innovations in solar and wind power have not only reduced our dependence on fossil fuels but also paved the way for a more sustainable future. These examples demonstrate how innovation has the power to reshape industries, disrupt existing paradigms, and create new opportunities.

While innovation brings about tremendous benefits, it also raises ethical considerations that need to be addressed. As technology advances, we must grapple with questions of privacy, data security, and the potential for misuse. The development of powerful algorithms and artificial intelligence systems raises concerns about bias, discrimination, and the impact on employment.

Ethical considerations also arise in fields such as genetic engineering and biotechnology. As we gain the ability to manipulate genes and potentially engineer life itself, we must carefully consider the ethical implications and ensure responsible use of these technologies.

To navigate these ethical challenges, it is essential to have open and inclusive dialogues that involve various stakeholders, including policymakers, scientists, and the general public. By actively engaging in discussions and establishing ethical frameworks, we can ensure that innovation is aligned with our shared values and serves the greater good.

Cultivating a culture of innovation within organisations is essential to drive progress and stay ahead in a rapidly evolving world. Here are some key strategies to foster innovation:

Encourage a growth mindset: Foster an environment that values continuous learning, encourages experimentation, and embraces failure as an opportunity for growth.

Promote diversity and inclusion: Embrace diverse perspectives and create an inclusive environment where everyone feels valued and empowered to contribute their unique insights.

Provide resources and support: Invest in training, infrastructure, and tools that enable employees to explore new ideas and turn them into reality.

Foster collaboration and cross-disciplinary work: Encourage collaboration across different teams, departments, and disciplines to foster knowledge sharing and spark innovation.

Recognise and reward innovation: Celebrate and acknowledge innovative ideas and efforts, creating a culture that values and incentivises innovation.

By implementing these strategies, organisations can create an environment that nurtures innovation, empowers employees, and drives meaningful progress.

Innovation is crucial for organisations to stay competitive. One of the key areas where innovation plays a pivotal role is new product development. The ability to create ground-breaking products that meet customer needs and desires is what sets successful companies apart. And now, with the advent of artificial intelligence (AI), the possibilities for innovation in new product development are expanding like never before.

AI has the potential to revolutionise the entire product development process, from ideation to market launch. By leveraging AI technologies such as machine learning and natural language processing, companies can gain valuable insights, generate innovative ideas, and optimise product designs. Management must explore AI use cases in new product development and consider how to implement the power of AI-driven innovation.

The integration of AI into the new product development process brings a multitude of benefits. First, AI enables companies to analyse vast amounts of data quickly and accurately. By feeding AI algorithms with market research data, consumer insights, and competitor analysis, organisations can gain deep and actionable insights that help them understand customer needs, preferences, and pain points. This knowledge forms the foundation for creating products that truly resonate with the target audience.

Second, AI can significantly speed up the idea generation and concept testing phases of new product development. With AI-powered tools, companies can automate the ideation process, generating a wide range of creative ideas based on predefined criteria. These tools can also evaluate the feasibility and potential success of these ideas, saving time and resources that would have otherwise been spent on manual evaluations.

Furthermore, AI can enhance the efficiency and accuracy of prototype development and testing. By simulating real-world scenarios and conducting virtual tests, companies can identify flaws and areas for improvement in their product designs before investing in physical prototypes. This not only saves costs but also reduces time-to-market, giving organisations a competitive edge in the fast-paced business environment.

AI use cases in market research and consumer insights: Market research and consumer insights are crucial for understanding the target market and identifying opportunities for innovation. AI can revolutionise these areas by automating data collection, analysis, and interpretation. Natural language processing algorithms can analyse customer reviews, social media posts, and other textual data to extract valuable insights about customer preferences, sentiments, and trends.

AI-powered chatbots and virtual assistants can also play a significant role in market research. By engaging with customers in real-time, these chatbots can gather feedback, answer questions, and collect valuable data about customer needs and pain points. This direct interaction with customers provides organisations with real-time insights that can inform their product development strategies.

Additionally, AI can assist in competitor analysis and benchmarking. Machine learning algorithms can analyse competitor data, pricing strategies, and product features to identify gaps in the market. This information can guide companies in developing innovative products that fill those gaps and offer a unique value proposition to customers.

AI use cases in idea generation and concept testing: Ideation is a critical phase in new product development, but it can be a time-consuming and challenging process. AI can streamline and enhance this phase by providing automated idea generation tools. These tools use AI algorithms to analyse existing product data, market trends, and customer preferences to generate a wide range of creative ideas.

Moreover, AI can assist in concept testing by simulating customer feedback and preferences. By analysing historical customer data, AI algorithms can predict how customers are likely to respond to different product concepts. This allows organisations to make data-driven decisions on which concepts to pursue further, saving time and resources that would have been spent on developing concepts that may not have resonated with the target market.

AI can also play a role in collaborative ideation and concept testing. Using AI-powered platforms, teams can collaborate remotely and in real-time, leveraging AI algorithms to generate ideas collectively. This can lead to more diverse and innovative concepts, as different perspectives and expertise are brought together.

AI use cases in prototype development and testing: Prototyping is a crucial phase in new product development, as it allows organisations to test and refine their product designs before investing in mass production. AI can enhance this phase by enabling virtual prototyping and testing.

With AI-powered tools, companies can create virtual prototypes that simulate real-world conditions and interactions. By analysing data from sensors and user interactions, these virtual prototypes can provide valuable insights into the usability, functionality, and ergonomics of the product. This allows organisations to identify design flaws and make iterative improvements at an early stage, saving costs and ensuring a better end product.

AI can also assist in user testing and feedback collection. By leveraging natural language processing, AI algorithms can analyse user feedback, reviews, and surveys to extract valuable insights. This feedback can guide organisations in refining their product designs and addressing user concerns, ensuring a product that truly meets customer needs.

AI use cases in product optimisation and improvement: Even after a product has been launched in the market, AI can continue to play a vital role in its optimisation and improvement. By analysing user behaviour data, AI algorithms can identify patterns, preferences, and pain points that can inform product enhancements.

AI-powered recommendation systems can also personalise the user experience by suggesting relevant products, features, or content based on individual preferences and behaviour. This not only enhances customer satisfaction but also drives repeat purchases and builds brand loyalty.

Furthermore, AI can assist in predictive maintenance and quality control. By analysing sensor data from connected devices, AI algorithms can detect anomalies, predict failures, and proactively initiate maintenance actions. This ensures that the product performs at its best and minimises downtime for the user.

While AI brings significant benefits to new product development, it is not without its challenges and limitations. One of the major challenges is the availability and quality of data. AI algorithms heavily rely on data to provide accurate insights and predictions. If the data used is incomplete, biased, or of poor quality, it can lead to inaccurate results and flawed decisions.

Another challenge is the need for expertise in AI technologies. Developing and implementing AI solutions require specialised skills and knowledge. Organisations need to invest in training their teams or collaborate with AI experts to ensure the successful integration of AI into the new product development process.

Moreover, ethical considerations and privacy concerns arise when using AI in new product development. Organisations need to ensure that the data is collected and analysis is done in a responsible and transparent manner, respecting user privacy and complying with relevant regulations.

A clear limitation is the inherent bias that AI algorithms may possess. If the training data used to develop AI models is biased, it can result in biased decision-making during the product development process. Another limitation is the lack of human intuition and creativity that AI currently possesses. While AI can analyse data and make predictions, it may struggle to understand complex human emotions or generate truly unique product ideas. Therefore, it is essential to strike a balance between AI capabilities and human expertise in product development.

Despite the numerous benefits of AI, there are several concerns that arise when incorporating AI use cases in new product development. These concerns must be addressed to ensure ethical, secure, and legally compliant product development processes.

Privacy and Security Concerns: One prominent concern with AI use cases in product development is the potential privacy and security risks associated with handling large volumes of sensitive data. AI relies heavily on data processing and analysis, and this often involves collecting and storing vast amounts of user information. Companies must prioritise data privacy and implement robust security measures to protect this information from unauthorised access or breaches. Transparency in data handling practices and obtaining user consent for data usage are crucial steps in addressing privacy concerns.

Ethical Considerations in AI Use Cases: Ethics play a crucial role in the use of AI in new product development. AI algorithms can inadvertently perpetuate biases present in the training data, leading to unfair or discriminatory outcomes. It is essential to ensure that AI models are trained on diverse and representative datasets to avoid biased decision-making in product development. Moreover, AI should not be used to manipulate or deceive users, and its use should align with ethical guidelines and industry standards. Building ethical frameworks and incorporating diverse perspectives in AI development can help address these concerns.

Legal and Regulatory Challenges: The increased use of AI in product development has raised various legal and regulatory challenges. Intellectual property rights, data protection regulations, and liability issues are some of the key concerns. Companies must ensure that they comply with relevant laws and regulations when using AI in product development. Additionally, clear guidelines and standards should be established to address potential liability issues that may arise from AI-driven decisions. Collaborating with legal experts and staying abreast of evolving regulations can help navigate these challenges effectively.

To address privacy and security concerns, companies should adopt a privacy-by-design approach when implementing AI in product development. This involves integrating privacy and security measures throughout the entire development process, rather than as an afterthought. Implementing robust data encryption, anonymisation techniques, and access controls can help mitigate security risks. Additionally, conducting regular security audits and ensuring compliance with data protection regulations are crucial steps in overcoming privacy and security concerns.

Ethical considerations can be addressed by implementing diverse and inclusive AI development practices. Ensuring that AI models are trained on representative datasets encompassing various demographics can help minimise biases. Companies should also establish clear guidelines and standards for AI development that prioritise fairness, transparency, and accountability. Regularly evaluating AI models for potential biases and involving ethics experts in the product development process can further enhance ethical considerations.

To overcome legal and regulatory challenges, companies should proactively collaborate with legal experts to ensure compliance with relevant laws. Conducting thorough due diligence regarding intellectual property rights and data protection regulations is essential before integrating AI into product development processes. Establishing internal policies and procedures for legal compliance and staying updated on evolving regulations can help mitigate legal and regulatory concerns effectively.

It is essential to clearly define the objectives and desired outcomes of using AI. This helps in selecting the right AI technologies and designing effective implementation strategies.

Organisations should invest in data quality and data governance. Data should be clean, accurate, and representative of the target market. This requires establishing robust data collection processes, data cleaning procedures, and data security measures. Furthermore, organisations should foster a culture of collaboration and learning. AI implementation is an ongoing process that requires continuous learning and adaptation. Encouraging cross-functional collaboration and providing training opportunities can help teams embrace AI technologies and leverage them effectively. Lastly, organisations should start small and iterate. Implementing AI in new product development can be complex and resource intensive. Starting with small-scale projects allows organisations to learn from the experience, identify challenges, and refine their AI strategies before scaling up.

While concerns with AI use cases in product development exist, numerous successful implementations demonstrate the potential of AI in driving innovation and improving product development processes. We can explore a couple of case studies that highlight the positive impact of AI.

Real-world case studies demonstrate the power of AI-driven innovation in new product development. One such example is the use of AI by a leading consumer goods company to optimise its product packaging design. By analysing customer feedback and preferences using AI algorithms, the company was able to identify the most appealing packaging design elements, leading to a significant increase in sales and customer satisfaction.

Another case study involves a tech startup that used AI-powered chatbots to collect real-time customer feedback during the product development phase. This direct interaction with customers helped the startup refine its product features and user experience, resulting in a highly successful product launch and a loyal customer base.

These case studies highlight the possibilities and benefits of using AI in new product development. They demonstrate how AI can provide valuable insights, enhance decision-making, and drive innovation, ultimately leading to successful product launches and business growth.

Toyota - Optimising Supply Chain Management: Toyota, a global manufacturing company, leveraged AI to optimise its supply chain management. By analysing historical data, AI algorithms identified patterns and trends, enabling accurate demand forecasting. This resulted in reduced inventory costs and enhanced supply chain efficiency. Moreover, AI-driven predictive maintenance algorithms helped identify potential machine failures in advance, minimising production downtime. Toyota's successful implementation of AI in its product development process showcases the benefits of AI in streamlining operations and improving overall productivity.

Amazon - Personalised Customer Recommendations: Amazon, an e-commerce giant, utilised AI to provide personalised customer recommendations. By analysing vast amounts of customer data, AI algorithms identified individual preferences and purchase patterns. This enabled the delivery of targeted product recommendations, enhancing the customer experience and driving sales. Amazon's AI-driven approach to new product development demonstrates the power of AI in understanding customer needs and tailoring products to meet those needs effectively.

As technology continues to advance, AI will undoubtedly play an increasingly important role in new product development. The ability to analyse vast amounts of data, generate innovative ideas, and optimise product designs makes AI a powerful tool for unlocking innovation.

However, it is important to recognise that AI is not a magic solution. Successful implementation requires careful planning, data quality, expertise, and ethical considerations. Organisations that embrace AI and leverage its capabilities while addressing these challenges will gain a competitive advantage in the dynamic business landscape.

The future of AI in new product development holds immense potential. As AI technologies continue to evolve, we can expect even more sophisticated and powerful tools that will further enhance innovation in product development. By harnessing the power of AI, organisations can unlock new possibilities, improve customer experiences, and drive business success.

https://orcid.org/0009-0005-0854-6213

CHAPTER 7

IT Systems and Management Information (MI)

ORGANISATIONS are leveraging artificial intelligence (AI) technologies to gain valuable insights, optimise processes, and make data-driven decisions. AI use cases in management information (MI) have the potential to reshape how businesses operate, enabling them to stay competitive in today's fast-paced environment.

Strategic management information is crucial for organisations to plan, execute, and evaluate their business strategies effectively. AI plays a pivotal role in transforming this information by analysing vast amounts of data, identifying patterns, and generating actionable insights. With AI, organisations can automate data collection, analysis, and reporting processes, saving valuable time and resources. Furthermore, AI algorithms can detect anomalies and predict future trends, empowering businesses to make informed decisions and stay ahead of the curve.

Financial management: In the realm of financial management, AI has proven to be a game-changer. AI-powered systems can analyse financial data, such as income statements, balance sheets, and cash flows, to identify potential risks and opportunities. These systems can also automate routine financial tasks, such as invoice processing and

DOI: 10.1201/9781003502708-8

expense management, freeing up valuable resources for more strategic initiatives. Moreover, AI algorithms can provide accurate financial forecasts, enabling organisations to make data-driven decisions and allocate resources effectively.

Supply chain management: AI is making a significant impact in supply chain management. AI technologies can optimise supply chain processes by analysing historical data, predicting demand fluctuations, and identifying bottlenecks. For instance, AI-powered demand forecasting systems can help organisations optimise inventory levels, reducing stockouts and excess inventory. Additionally, AI algorithms can optimise delivery routes, minimising transportation costs and improving overall efficiency.

Human resources management: AI is revolutionising human resources (HR) management by automating mundane HR tasks and improving decision-making processes. For instance, AI-powered chatbots can handle employee enquiries, providing instant responses and freeing up HR professionals' time. AI algorithms can also analyse employee data to identify performance patterns, predict attrition, and recommend personalised development plans. Furthermore, AI can assist in unbiased hiring decisions by removing unconscious biases from the recruitment process.

Customer relationship management (CRM): AI is transforming how organisations manage their customer relationships. AI-powered CRM systems can analyse customer data, such as purchase history and online behaviour, to personalise marketing campaigns and improve customer satisfaction. AI algorithms can also automate customer support processes, enabling organisations to provide round-the-clock assistance. By leveraging AI in CRM, businesses can enhance customer engagement, increase loyalty, and drive revenue growth.

Risk management: Risk management is a critical aspect of strategy and management. AI can enhance risk management processes by analysing vast amounts of data and identifying potential risks and vulnerabilities. AI algorithms can analyse historical data, market trends, and external factors to identify patterns and anomalies that may indicate potential risks.

By leveraging AI-powered risk management tools, organisations can automate the process of risk identification, assessment, and mitigation. This enables organisations to proactively manage risks, make informed decisions, and minimise the impact of potential disruptions.

Moreover, AI can assist in real-time risk monitoring by analysing data from various sources, such as social media, news articles, and financial reports, to detect emerging risks and enable timely interventions. AI-powered risk management systems can analyse vast amounts of data to identify potential risks, detect fraud, and predict market trends. By leveraging AI in risk management, organisations can proactively mitigate potential threats and safeguard their operations.

Project management: AI is transforming project management by automating repetitive tasks, improving collaboration, and enhancing decision-making. AI-powered project management systems can automate project scheduling, resource allocation, and progress tracking. These systems can also analyse project data, such as task completion times and resource utilisation, to identify inefficiencies and optimise project performance. Additionally, AI algorithms can assist project managers in making data-driven decisions by providing insights and recommendations based on historical data.

Data analysis and decision-making: Data analysis and decision-making are at the core of effective MI. AI technologies can analyse large datasets, identify patterns, and generate valuable insights. For example, AI algorithms can perform sentiment analysis on customer feedback, enabling organisations to understand customer preferences and improve their products or services. AI-powered decision support systems can also provide recommendations based on historical data, helping organisations make informed decisions and drive business success.

Strategic planning: Strategic planning is a critical aspect of any organisation's success. AI can play a pivotal role in this process by analysing large volumes of data, identifying trends, and generating insights. With AI-powered tools, organisations can conduct comprehensive market research, assess competitive landscapes, and identify emerging opportunities. These insights enable strategic decision-making based on data-driven intelligence, giving organisations a competitive edge.

Furthermore, AI can assist in scenario planning and predictive modelling. By analysing historical data and market trends, AI algorithms can simulate different scenarios, enabling organisations to evaluate the potential outcomes of various strategic choices. This empowers decision-makers to make informed decisions and develop robust strategic plans.

Performance management: Performance management is crucial for organisations to monitor, evaluate, and improve employee performance. AI can enhance this process by leveraging machine learning algorithms to analyse vast amounts of data, including employee performance metrics, feedback, and historical data. By doing so, AI can identify patterns and trends that may not be apparent to human managers. This enables organisations to provide personalised feedback, identify skill gaps, and develop targeted training programmes to improve employee performance and productivity.

Moreover, AI-powered performance management tools can automate time-consuming administrative tasks, such as performance evaluations and goal setting. This frees up valuable time for managers to focus on more strategic initiatives, fostering a culture of continuous improvement within the organisation.

Decision-making: Effective decision-making is at the core of successful organisations. AI can enhance decision-making processes by analysing vast amounts of data and extracting relevant insights. By leveraging machine learning algorithms, AI can process data faster and more accurately than humans, enabling organisations to make data-driven decisions in real-time.

AI can also assist in decision-making by providing predictive analytics. By analysing historical data and market trends, AI algorithms can forecast future scenarios, enabling organisations to make proactive decisions and stay ahead of the competition. Additionally, AI can identify patterns and anomalies in data, helping organisations detect potential risks and opportunities that may have otherwise gone unnoticed.

Resource allocation: Efficient resource allocation is essential for organisations to optimise productivity and achieve strategic objectives. AI can assist in this process by analysing data on resource utilisation, demand patterns, and market dynamics. By doing so, AI algorithms can identify areas of inefficiency and recommend optimal resource allocation strategies.

For example, in supply chain management, AI can analyse data on inventory levels, customer demand, and production capacity to optimise inventory management and reduce costs. AI can also help organisations allocate human resources effectively by analysing employee skills, workload, and project requirements. By leveraging AI-powered resource allocation tools, organisations can streamline operations, increase efficiency, and improve overall performance.

Implementing AI in strategy and management requires careful planning and execution. Organisations should follow a systematic approach to ensure successful implementation and maximise the benefits of AI.

Identify strategic objectives: Before implementing AI, organisations need to identify their strategic objectives and align AI initiatives accordingly. This involves understanding the specific challenges and opportunities in strategy and management and defining clear goals for AI implementation.

Data collection and preparation: AI relies on high-quality data for accurate analysis and insights. Organisations should invest in data collection and preparation processes to ensure they have access to clean, structured, and relevant data. This may involve integrating data from various sources, implementing data governance practices, and ensuring data security and privacy.

Selecting the right AI tools and technologies: There is a wide range of AI tools and technologies available in the market. Organisations should carefully evaluate their requirements and select the most suitable tools that align with their strategic objectives. This may involve considering factors such as scalability, ease of integration, and user-friendliness.

Training and integration: Once the AI tools and technologies are selected, organisations need to provide adequate training to employees to ensure they can effectively utilise AI capabilities. This may involve upskilling employees, providing training programmes, and fostering a culture of continuous learning.

Monitoring and evaluation: Implementing AI is an iterative process that requires continuous monitoring and evaluation. Organisations should establish key performance indicators (KPIs) to measure the effectiveness of AI initiatives and make necessary adjustments. Regular monitoring and evaluation enable organisations to identify areas of improvement and optimise AI implementation.

While AI offers immense potential in strategy and management, there are several challenges and considerations that organisations need to address.

Ethical considerations: AI raises ethical concerns related to privacy, bias, and transparency. Organisations need to ensure that AI algorithms and models are fair, unbiased, and transparent. This involves addressing issues such as algorithmic bias, data privacy, and explainability of AI decisions.

Data quality and security: AI relies on high-quality data for accurate analysis and insights. Organisations need to invest in data quality assurance processes to ensure they have access to clean and reliable data. Additionally, organisations need to implement robust data security measures to protect sensitive information from unauthorised access or breaches.

Employee adoption and resistance: Implementing AI may face resistance from employees who fear job displacement or feel overwhelmed by new technologies. Organisations need to address these concerns by providing adequate training and support to employees, emphasising the augmentation rather than replacement of human capabilities.

Regulatory and legal compliance: AI implementation should comply with relevant regulatory and legal frameworks. Organisations need to ensure that their AI initiatives adhere to data protection laws, intellectual property rights, and other legal requirements.

The field of AI is continuously evolving, and there are several future trends that will have a profound impact on strategy and management.

Natural language processing (NLP) and conversational AI: NLP enables machines to understand and interpret human language. NLP-powered conversational AI interfaces will enable organisations to interact with AI systems using natural language, making AI more accessible and user-friendly.

Explainable AI: Explainable AI aims to make AI decisions more transparent and understandable. Explainable AI algorithms provide explanations for their decisions, enabling organisations to gain insights into how AI arrived at a particular conclusion. This enhances trust in AI systems and enables organisations to identify and address biases or errors.

Augmented intelligence: Augmented intelligence refers to the collaboration between humans and AI systems to enhance decision-making and problem-solving. Rather than replacing humans, AI acts as a tool to augment human capabilities, enabling organisations to leverage the unique strengths of both humans and machines.

Edge computing and AI: Edge computing involves processing data at the edge of the network, closer to the data source. By combining edge computing with AI capabilities, organisations can leverage real-time data analysis and insights, enabling faster decision-making and reducing reliance on cloud-based infrastructure.

Ethical AI: The ethical considerations surrounding AI will become even more critical in the future. Organisations will need to develop ethical frameworks and guidelines for AI usage, ensuring that AI systems are fair, transparent, and accountable.

AI has the potential to revolutionise strategy and management processes. By leveraging AI in strategic planning, performance management, decision-making, resource allocation, and risk management, organisations can unlock a competitive edge and drive success. However, implementing AI requires careful planning, addressing ethical considerations, ensuring data quality and security, and fostering employee adoption. As AI continues to evolve, future trends such as NLP, explainable AI, augmented intelligence, edge computing, and ethical AI will shape the future of strategy and management.

Numerous organisations have already embraced AI in their strategic MI processes. For instance, a multinational retailer leverages AI algorithms to analyse customer buying patterns and optimise inventory levels across their global supply chain. This has resulted in reduced stockouts and improved customer satisfaction. Another example is a financial institution that uses AI-powered risk management systems to detect fraudulent activities and protect customer assets. These real-world examples demonstrate the tangible benefits of implementing AI in MI.

The field of AI in MI is continuously evolving, and several future trends and advancements are worth noting. One such trend is the increasing use of NLP technologies to analyse unstructured data, such as customer reviews and social media posts. This enables organisations to gain valuable insights from vast amounts of textual data. Additionally, advancements in machine learning algorithms and deep learning techniques are expected to further enhance the accuracy and efficiency of AI systems in MI.

AI use cases in MI have the potential to transform how organisations operate and make decisions. From financial management to supply chain optimisation, AI technologies are revolutionising various aspects of MI. By leveraging AI, organisations can gain valuable insights, automate routine tasks, and make data-driven decisions. As AI continues to advance, it is crucial for businesses to embrace these game-changing technologies to stay competitive in the digital age.

AI has rapidly transformed industries. While AI offers immense potential for innovation and efficiency, it also raises significant ethical concerns. As technology evolves, it becomes crucial to examine the moral dilemmas associated with AI use cases in the IT sector. Ethical considerations should be at the forefront of decision-making to ensure that AI is developed and implemented responsibly.

One of the primary ethical concerns with AI in IT is the potential invasion of privacy and data security breaches. AI systems often require access to vast amounts of personal and sensitive data to function effectively. However, this raises questions about how this data is collected, stored, and protected. Organisations must prioritise robust data protection measures, implement stringent security protocols, and obtain explicit consent from individuals to avoid compromising their privacy.

Before we dive into the specific use cases of AI in IT, it is important to have a clear understanding of what AI is and how it can be applied in the industry. AI refers to the simulation of human intelligence in machines that are programmed to think and learn like humans. In the IT industry, AI can be used to automate repetitive tasks, analyse vast amounts of data, and make intelligent decisions.

AI use cases in cybersecurity: In the ever-evolving landscape of cybersecurity, AI has become an indispensable tool for detecting and preventing cyber threats. AI algorithms can analyse large volumes of data to identify patterns and anomalies that may indicate a potential security breach. By continuously learning from new data, AI systems can adapt and improve their ability to detect and respond to emerging threats in real-time. This proactive approach to cybersecurity is crucial in safeguarding sensitive data and preventing potential cyber-attacks.

AI-powered cybersecurity solutions can also help in fraud detection and prevention. By analysing transactional data and user behaviour patterns, AI algorithms can detect fraudulent activities and alert IT teams to take appropriate action. This not only helps in protecting businesses from financial losses but also enhances customer trust and confidence.

Enhancing IT support with AI: One of the most significant use cases of AI in IT is in the field of IT support and helpdesk operations. AI-powered chatbots and virtual assistants can provide instant and personalised assistance to users, reducing the need for human intervention. These AI-driven systems can handle common support queries, troubleshoot technical issues, and even provide step-by-step guidance for problem resolution.

By leveraging NLP and machine learning algorithms, AI chatbots can understand and respond to user queries in a conversational manner. This not only improves the overall user experience but also frees up IT support teams to focus on more complex and critical tasks. Additionally, AI systems can learn from user interactions and continuously improve their knowledge base, ensuring accurate and efficient support.

AI in data management and analysis: The amount of data generated by IT systems is growing exponentially, making it increasingly challenging for organisations to manage and analyse this vast amount of information. AI can play a crucial role in addressing this challenge by automating data management processes and facilitating advanced data analysis.

AI-powered data management systems can automatically classify and tag data, making it easier to organise and retrieve information when needed. Moreover, AI algorithms can analyse large datasets to identify trends, patterns, and correlations that may not be apparent to human analysts. This enables organisations to gain valuable insights from their data and make data-driven decisions.

AI-powered automation in IT operations: IT operations involve a wide range of routine tasks that can be time-consuming and prone to human error. AI-powered automation can streamline these operations and improve overall efficiency. By automating repetitive tasks such as system monitoring, patch management, and software updates, AI systems can free up IT professionals to focus on more strategic and value-added activities.

AI algorithms can monitor IT systems in real-time, detect anomalies, and automatically initiate remedial actions. This proactive approach helps in identifying and resolving potential issues before they impact the business. Moreover, AI systems can learn from historical data and make predictions about future system behaviour, enabling organisations to take preventive measures and optimise system performance.

AI for predictive maintenance and system optimisation: Predictive maintenance is another area where AI can have a significant impact on IT. By analysing historical performance data and real-time sensor readings, AI algorithms can predict when IT systems are likely to fail or require maintenance. This enables organisations to schedule maintenance activities proactively, minimising downtime and reducing costs associated with unexpected system failures.

AI systems can also optimise IT infrastructure by dynamically allocating resources based on workload demands. By analysing patterns and trends in system usage, AI algorithms can predict resource requirements and automatically allocate resources accordingly. This ensures that IT systems operate at optimal efficiency while minimising costs and improving overall performance.

AI-driven chatbots for customer service in IT: In the digital era, customer service has become a critical aspect of IT operations. AI-driven chatbots have emerged as a valuable tool for providing efficient and personalised customer support. These chatbots can handle a wide range of customer queries, provide instant responses, and guide users through troubleshooting processes.

AI chatbots can be integrated with IT service management systems, allowing them to access relevant information and provide accurate and timely responses. Moreover, AI algorithms can analyse customer interactions to identify common issues and trends, enabling organisations to proactively address customer concerns and improve overall customer satisfaction.

Several organisations have successfully implemented AI in their IT operations, reaping significant benefits. For instance, a global e-commerce company implemented AI-powered fraud detection algorithms, which helped them detect and prevent fraudulent transactions in real-time. This resulted in substantial cost savings and increased customer trust.

Another real-life example is a multinational IT services company that deployed AI chatbots for IT support. These chatbots were able to handle 80% of customer queries, reducing the workload on IT support teams and improving response times. This led to enhanced customer satisfaction and improved overall efficiency.

An ethical concern that arises with AI use cases in IT is the issue of bias and discrimination. AI algorithms are designed to learn from existing data, which can inadvertently perpetuate biases present in the data. This can result in discriminatory outcomes, such as biased hiring practices or unfair treatment of certain individuals or groups. It is crucial to address these biases during the development phase of AI systems by implementing diverse and representative datasets and conducting rigorous testing to identify and rectify any biases.

Transparency and explainability are also critical ethical considerations in AI decision-making. AI algorithms often make complex decisions that impact individuals' lives, such as credit scoring or medical diagnoses. However, these decisions are often considered black boxes, where it is challenging to understand how the AI arrived at a particular outcome. This lack of transparency can erode trust in AI systems and lead to unjust outcomes. Organisations must prioritise developing AI systems that are transparent and provide explanations for their decisions, allowing individuals to understand and challenge them if needed.

Privacy issues and data security: Responsible AI implementation also demands a focus on responsibility and accountability. As AI systems become more autonomous and make decisions independently, it becomes essential to define who is responsible when things go wrong. The IT department must establish clear lines of accountability and ensure that individuals are held responsible for the actions and decisions of AI systems under their supervision. This includes implementing mechanisms for auditing AI systems and providing avenues for redress in case of any harm caused by AI.

Regulatory and legal challenges further complicate the ethical landscape of AI use cases in the IT department. The rapid pace of AI development often outpaces the formulation of comprehensive regulations. This creates a legal grey area where organisations struggle to navigate the ethical implications of AI use. Policymakers must work closely with industry experts to develop robust regulations that address the unique challenges posed by AI systems, ensuring that ethical considerations are adequately addressed and enforced.

Bias and discrimination: Mitigating ethical concerns in AI use cases requires a multi-faceted approach. Organisations should prioritise diversity and inclusion in the development teams responsible for creating AI systems. A diverse team brings different perspectives and experiences, reducing the risk of bias and discrimination in AI algorithms. Furthermore, organisations must invest in ongoing training and education to ensure that employees understand the ethical implications of AI and are equipped to make responsible decisions.

Regular audits and assessments of AI systems should be conducted to identify and rectify any biases or discriminatory patterns. This includes monitoring the performance and impact of AI algorithms and making necessary adjustments to ensure fairness and equity. Additionally, organisations should actively engage with external stakeholders, such as ethicists and advocacy groups, to gain diverse perspectives and insights on the ethical implications of AI use cases in IT.

Transparency and explainability: Striking the right balance between innovation and ethics is paramount when it comes to AI use cases in the IT department. While AI offers immense potential for improved efficiency and decision-making, it also presents ethical concerns that must not be overlooked. Privacy issues, bias and discrimination, transparency, responsibility, and legal challenges all demand careful consideration and proactive measures to ensure that AI is deployed responsibly and ethically.

Organisations that prioritise ethical considerations in AI use cases will not only mitigate potential harm but also gain public trust and confidence in their AI systems. As AI continues to evolve, it is essential to foster a culture of responsibility and accountability within IT and the wider organisation. By doing so, we can harness the power of AI while upholding ethical standards and ensuring a more equitable and just future.

While the potential benefits of AI in IT are undeniable, there are several challenges and considerations that organisations need to address when adopting AI technologies. One of the key challenges is the availability of high-quality data. AI algorithms rely on large volumes of high-quality data to learn and make accurate predictions. Organisations need to ensure that they have access to clean and relevant data to train their AI systems effectively.

Another challenge is the ethical and legal implications of AI. As AI systems become more autonomous and make decisions that impact humans, it is crucial to establish ethical guidelines and regulations to ensure fairness and accountability. Organisations must also consider the potential impact of AI on the workforce and proactively address any concerns related to job displacement.

AI has the potential to revolutionise the IT industry by automating routine tasks, improving cybersecurity, enhancing customer support, and optimising IT operations. Real-life examples have shown the significant benefits that organisations can achieve by embracing AI in their IT operations. However, it is important to address the challenges and considerations associated with AI adoption to ensure successful implementation. By harnessing the power of AI, organisations can unlock new possibilities and pave the way for a brighter future in IT.

https://orcid.org/0009-0005-0854-6213

CHAPTER 8

Individual Liberties

EMERGING TECHNOLOGIES have transformed the way we live, work, and interact with the world. From artificial intelligence to biotechnology, these advancements hold immense promise for the future. However, they also raise significant ethical concerns, particularly when it comes to the balance between technological progress and individual liberties. We must explore the ethical dilemmas posed by emerging technologies and the need to find a harmonious coexistence between innovation and the protection of individual rights.

To comprehend the ethical challenges associated with emerging technologies, it is crucial to first understand what these technologies entail. Emerging technologies refer to novel innovations that are rapidly advancing and reshaping various sectors. They encompass a wide range of fields, such as robotics, virtual reality, genetic engineering, and nanotechnology.

While these technologies offer immense potential for improving efficiency, enhancing healthcare, and addressing global challenges, they also raise concerns about individual liberties. The use of artificial intelligence, for example, has prompted debates about privacy invasion and algorithmic bias. Similarly, genetic engineering and biotechnology have sparked discussions about the ethical implications of manipulating human genes and the potential for discrimination based on genetic information.

The ethical debate surrounding emerging technologies centres on striking a balance between technological progress and the protection of individual rights and liberties. On the one hand, proponents argue that technological advancements have the power to uplift societies, improve quality of life, and solve complex problems. They advocate for an

DOI: 10.1201/9781003502708-9

environment that fosters innovation and allows emerging technologies to flourish.

On the other hand, critics emphasise the need to safeguard individual rights and ensure that technological advancements do not infringe upon human dignity, privacy, or autonomy. They raise concerns about the potential for abuse, discrimination, and unequal distribution of benefits that may arise from the unchecked development and deployment of emerging technologies.

The rapid pace of technological innovation has outpaced the development of regulatory frameworks to address the ethical concerns associated with emerging technologies. This has given rise to a multitude of concerns regarding the potential infringement on individual liberties.

One significant concern is the erosion of privacy. With the proliferation of surveillance technologies and the collection of vast amounts of personal data, individuals' privacy is increasingly at risk. The use of facial recognition technology, for instance, raises questions about the balance between security and the right to anonymity in public spaces.

Another concern revolves around the potential for discrimination. As emerging technologies become integrated into various aspects of our lives, there is a risk that they may perpetuate existing inequalities or create new forms of discrimination. For example, algorithms used in hiring processes may inadvertently favour certain demographic groups, leading to biased outcomes.

Furthermore, the ethical implications of emerging technologies extend to issues of consent and autonomy. For instance, advancements in medical technology raise questions about the extent to which individuals should have control over their own genetic information or the use of their personal health data.

To illustrate the ethical dilemmas posed by emerging technologies, let us examine a few real-world examples:

Autonomous vehicles: The development of self-driving cars raises questions about liability and ethical decision-making. In the event of an accident, who should be held responsible – the vehicle manufacturer, the software developer, or the individual in the car? Additionally, how should autonomous vehicles make ethical decisions in situations where harm is inevitable, such as choosing between hitting a pedestrian or swerving into oncoming traffic?

Facial recognition technology: The widespread use of facial recognition technology by law enforcement agencies raises concerns about privacy and potential abuse. The technology has the potential to infringe upon individuals' rights to anonymity and due process, as well as perpetuate systemic biases.

Genetic engineering: The ability to edit genes using technologies like CRISPR-Cas9 holds immense promise for treating genetic diseases. However, it also raises ethical questions about the boundaries of genetic manipulation, the potential for designer babies, and the long-term consequences of altering the human genome.

These case studies highlight the complex ethical dilemmas that emerge when emerging technologies intersect with individual liberties.

To address the ethical challenges posed by emerging technologies, various ethical frameworks have been proposed. These frameworks provide guiding principles for decision-making and ensure that individual liberties are protected.

One such framework is the principle of privacy by design, which advocates for incorporating privacy safeguards into the design and development of technologies from the outset. By embedding privacy into the technological infrastructure, individuals' rights to privacy are respected, and potential abuses are minimised.

Another ethical framework is the principle of transparency and explainability. This framework emphasises the importance of making the decision-making processes of emerging technologies transparent and understandable to individuals. By providing clear explanations of how algorithms work and the data they use, individuals can make informed choices and hold technology companies accountable.

Additionally, the principle of inclusive design promotes the idea that emerging technologies should be developed with diverse users in mind. By considering a wide range of perspectives and experiences, technologies can be designed to minimise bias and ensure equal access and benefits for all individuals.

Government and regulatory bodies play a crucial role in safeguarding individual rights in the face of emerging technologies. They have the authority to establish and enforce laws and regulations that protect privacy, prevent discrimination, and ensure accountability.

To effectively address the ethical challenges posed by emerging technologies, governments need to adopt a proactive approach. This involves conducting thorough risk assessments, engaging in public consultations, and collaborating with experts to develop comprehensive regulatory frameworks.

Furthermore, regulatory bodies should have the power to monitor and assess the compliance of technology companies with ethical standards. This includes conducting audits, imposing fines for non-compliance, and revoking licences when necessary.

It is essential for governments to strike a delicate balance between fostering innovation and protecting individual liberties, ensuring that emerging technologies serve the common good without infringing upon fundamental rights.

Technology companies also bear a significant responsibility in protecting individual liberties when developing and deploying emerging technologies. They have the power to shape the ethical landscape through their design choices, policies, and practices.

First and foremost, technology companies should prioritise privacy and data protection. They should adopt privacy by design principles, implement robust security measures, and provide users with clear and meaningful consent mechanisms.

Moreover, technology companies should actively address issues of algorithmic bias and discrimination. This involves regularly auditing and testing algorithms for biases, investing in diverse talent and perspectives, and ensuring transparency in algorithmic decision-making processes.

Additionally, technology companies should engage in ethical marketing practices and avoid exploiting individuals' personal data for profit. They should prioritise the well-being and autonomy of their users over commercial interests.

By taking these steps, technology companies can demonstrate their commitment to protecting individual liberties and contribute to a more ethical and responsible technological landscape.

As emerging technologies continue to advance at an unprecedented rate, it is important to anticipate and address future ethical challenges. This requires ongoing dialogue, collaboration, and proactive measures from all stakeholders involved – governments, regulatory bodies, technology companies, and individuals.

Some of the future ethical challenges that may arise include the ethical implications of brain–computer interfaces, the impact of augmented reality on privacy and consent, and the ethical considerations of artificial general intelligence.

To navigate these challenges, it is crucial for society to prioritise ethical considerations alongside technological advancements. By fostering an environment that values individual liberties, upholds ethical standards, and promotes inclusive decision-making, we can ensure that emerging technologies serve as a force for good and benefit humanity as a whole.

The ethical dilemmas posed by emerging technologies require us to find a delicate balance between progress and individual liberties. While these technologies hold immense promise, they also raise concerns about privacy, discrimination, and consent.

By understanding the impact of emerging technologies on individual liberties, adopting ethical frameworks, and involving all stakeholders in the decision-making process, we can navigate the ethical challenges presented by these advancements. Governments, regulatory bodies, and technology companies must work together to establish comprehensive regulations, protect privacy, and ensure accountability.

Ultimately, by striving to strike a balance between emerging technologies and individual liberties, we can harness the potential of these innovations while upholding the fundamental rights and values that define our society.

https://orcid.org/0009-0005-0854-6213

CHAPTER 9

Autonomy, Authenticity, and Identity

EMERGING TECHNOLOGIES have reformed the way we live, work, and interact with the world. From artificial intelligence to virtual reality, these advancements offer incredible opportunities for growth and progress. However, with every new development comes concerns about the impact on our autonomy, authenticity, and identity. Societies must explore these concerns and discuss strategies for navigating them in the age of emerging technologies.

One of the primary concerns surrounding emerging technologies is the potential loss of autonomy. As these technologies become more integrated into our daily lives, there is a fear that they may infringe upon our ability to make independent decisions. For example, autonomous vehicles raise questions about who is responsible in the event of an accident. Are the passengers at fault or the technology itself?

To navigate these concerns, it is crucial to establish robust regulations and ethical frameworks. Governments should work closely with technology developers to create guidelines that prioritise individual autonomy while still harnessing the benefits of these technologies. Additionally, individuals should educate themselves about the potential risks and limitations of emerging technologies to make informed choices about their usage.

DOI: 10.1201/9781003502708-10

Another concern that arises in the age of emerging technologies is the question of authenticity. As virtual reality and augmented reality become more prevalent, there is a worry that our experiences may become diluted or artificial. For instance, online interactions can lack the genuine emotions and connections that we experience in face-to-face interactions.

To address these concerns, it is essential to strike a balance between the virtual and physical worlds. While emerging technologies offer exciting possibilities, it is crucial to maintain a connection to our authentic selves and the world around us. Prioritising real-life experiences, fostering genuine relationships, and being mindful of the impact of technology on our well-being can help us preserve our authenticity in this rapidly evolving digital landscape.

The emergence of new technologies also raises concerns about our identity. With the rise of social media and virtual identities, there is a risk of losing sight of our true selves. The pressure to curate a perfect online persona can lead to feelings of insecurity and a distorted sense of identity. Additionally, the collection and analysis of personal data by emerging technologies can pose risks to our privacy and personal security.

To navigate these concerns, it is important to cultivate self-awareness and a strong sense of identity. We should strive to maintain a healthy balance between our online and offline lives, ensuring that our virtual identities align with our authentic selves. Taking control of our personal data and being mindful of the information we share can also help protect our identity in the digital realm.

While concerns about autonomy, authenticity, and identity are valid, it is crucial to recognise that emerging technologies also bring significant benefits. They have the potential to improve our quality of life, enhance productivity, and advance societal progress. The key lies in striking a balance between embracing technological advancements and staying true to our personal values.

To achieve this balance, we must reflect on our values and priorities. What matters most to us? How do we want technology to enhance our lives without compromising our autonomy, authenticity, and identity? By aligning our technological choices with our values, we can navigate the concerns surrounding emerging technologies with confidence and integrity.

Navigating the concerns of autonomy, authenticity, and identity in the age of emerging technologies requires a proactive approach. Here are some strategies to help you navigate these concerns:

Raising awareness: Helping raise awareness of the ethical concerns can help foster dialogue between stakeholders such as government, regulators, scientists, and researchers to consider the latest technological advancements and their potential implications. Education on the risks and benefits can help foster more informed decisions about their usage.

Setting boundaries: Establishing frameworks and clear boundaries for the use of emerging technologies can help safe research to proceed whilst setting guardrails where risks are posed.

Fostering human connections: Emphasising the importance of human, face-to-face interaction, meaningful relationships, and open conversations can help sway the future direction of technological advancements. Authenticity, human connection, and human intervention can counterbalance the potential artificiality and misinterpretation of emerging technologies.

Protecting the privacy of individuals: Agreements over how personal information can or cannot be used can help protect the privacy of individuals.

Reflection and discussion over societal values: Societies must consider the use of emerging technologies and their values to prevent the erosion of their society. Regulators must consider how technology may enhance the lives of citizens whilst agreeing guardrails to preserve autonomy and authenticity.

By implementing these strategies, you can navigate the concerns of autonomy, authenticity, and identity in the age of emerging technologies while embracing the benefits they offer.

The age of emerging technologies presents us with exciting opportunities for growth and progress. However, concerns surrounding autonomy, authenticity, and identity must be addressed to ensure a harmonious integration of technology into our lives. By understanding these concerns and implementing strategies for navigation, societies can embrace the benefits of emerging technologies while staying true to ourselves.

https://orcid.org/0009-0005-0854-6213

CHAPTER 10

Playing God and Hubris

EMERGING TECHNOLOGIES have always been at the forefront of human progress. They hold the promise of transforming our lives, revolutionising industries, and solving complex problems. From artificial intelligence to genetic engineering, these emerging technologies are poised to shape the future in ways we can only imagine. They offer the potential to improve healthcare, enhance communication, increase efficiency, and create new opportunities for growth and innovation.

One of the most exciting promises of emerging technologies is their ability to improve healthcare outcomes. With advancements in medical technology, we can expect to see the most accurate diagnoses, personalised treatments, and improved patient care. For example, general editing technologies like CRISPR have the potential to eradicate genetic diseases by modifying the DNA of living organisms. Similarly, artificial intelligence-powered algorithms can analyse vast amounts of medical data to detect patterns, predict diseases, and recommend personalised treatments.

Another promise of emerging technologies lies in their ability to enhance communication connectivity. The rise of the internet and mobile devices has already transformed the way people communicate, but emerging technologies take it a step further. Virtual reality and augment reality technologies are revolutionising the way we interact with each other and our surroundings. They enable us to have immersive experiences, collaborate remotely, and access information in unprecedented ways. As these technologies continue to advance, we expect even greater connectivity and integration into our daily lives.

DOI: 10.1201/9781003502708-11

While the promises of emerging technologies are undoubtedly exciting, it is crucial to recognise and address the potential perils they bring. One of the primary concerns surrounding emerging technologies is the ethical dilemma of playing God. As humans, we are now capable of manipulating the very building blocks of life, altering the course of evolution, and making decisions that impact the future of our species. This power raises questions about our responsibility and the potential consequences of playing God.

The ethical concerns surrounding playing God through emerging technologies are multi-faceted. First and foremost, there is the question of whether we should tamper with nature and the fundamental principles of life. Some argue that it is our duty to use these technologies to alleviate human suffering and improve the overall well-being of society. Others caution against the hubris of thinking that we can control nature and predict the long-term consequences of our actions.

Another concern is the potential for misuse and unintended consequences. As emerging technologies become more powerful and accessible, the risk of abuse and unintended harm increases. For example, genetic engineering technologies can be used for malevolent purposes, such as creating bioweapons or perpetuating genetic discrimination. Similarly, artificial intelligence-powered systems can inadvertently perpetuate biases and inequalities not carefully designed and regulated.

The concept of playing God has deep-rooted ethical implications that are closely tied to emerging technologies. By assuming the role of creators and manipulators of life, we are faced with profound moral questions. These questions revolve around the limits of our power, the potential consequences of our actions, and the responsibility we have towards the natural world and future generations.

One of the key ethical concerns is the potential loss of human dignity and autonomy. As we gain the ability to modify our genetic makeup and enhance our cognitive abilities, we risk devaluing the inherent worth of being human. The pursuit of perfection and the desire to transcend our natural limitations may come at the cost of our uniqueness and individuality. Additionally, the unequal distribution of these technologies can exacerbate existing inequalities and create a new form of social divide.

Another ethical concern is the potential disruption of natural ecosystems and the delicate balance of life. Emerging technologies like geoengineering, which aims to manipulate the Earth's climate systems, may have unintended consequences on biodiversity and ecological stability. Playing

God without a comprehensive understanding of the interconnectedness of life can have far-reaching and irreversible impacts on the planet and all its inhabitants.

While the concerns surrounding playing God through emerging technologies are valid, there are also potential benefits that should not be overlooked. These technologies have the potential to improve human health, enhance our understanding of the natural world, and address pressing global challenges.

In the field of healthcare, emerging technologies offer the promise of personalised medicine, targeted therapies, and regenerative medicine. Gene editing technologies like CRISPR can potentially cure genetic diseases by correcting faulty genes. Stem cell research and tissue engineering hold the potential to regenerate damaged organs and tissues, offering hope to those suffering from chronic diseases and injuries.

Moreover, playing God through emerging technologies can provide us with a deeper understanding of nature and the universe. For example, advancements in astronomy and exploration technologies enable us to explore distant galaxies, study celestial bodies, and unlock the mysteries of the cosmos. These discoveries not only expand our knowledge but also inspire wonder and curiosity, fuelling further scientific advancements.

Additionally, emerging technologies can help address pressing global challenges, such as climate change and food security. Sustainable energy technologies like solar and wind power can reduce our dependence on fossil fuels and mitigate the impact of climate change.

Agricultural technologies like precision farming and genetically modified crops can increase food production and ensure food security for a growing global population.

While the potential benefits of playing God through emerging technologies are vast, it is essential to recognise and mitigate the potential risks and dangers they pose. These risks range from unintended consequences and ethical dilemmas to the loss of human agency and the potential for misuse.

One of the primary risks is the unintended consequences of our actions. As we manipulate building blocks of life and alter natural systems, we may inadvertently disrupt the delicate balance of ecosystems and trigger cascading effects. For example, introducing genetically modified organisms into the environment may have unforeseen ecological consequences, impacting biodiversity and ecosystem stability.

Another risk is the ethical dilemmas that arise from playing God. As we gain the ability to modify our genetic makeup and enhance our cognitive abilities, we face difficult decision about what is morally acceptable and what crosses ethical boundaries. For example, using here genetic engineering technologies for non-therapeutic purposes, such as enhancing physical appearance or intelligence, raises questions about fairness, equality, and the commodification of human traits.

Furthermore, there is the risk of misuse and abuse of emerging technologies. As these technologies become more powerful and accessible, they can be weaponised or used for malevolent purposes. For example, the use of artificial intelligence in autonomous weapon systems raises concerns about the loss of human control and the potential for indiscriminate harm.

Emerging technologies have transformed our lives in countless ways, offering promising solutions to complex problems. However, as these technologies continue to advance at an unprecedented pace, there is a growing concern regarding the presence of hubris within the industry. Hubris, defined as excessive pride or self-confidence, can blind us to potential risks and ethical considerations. We can consider the concept of hubris within the world of emerging technologies, explore the common concerns surrounding these advancements, and discuss the importance of recognising and addressing these concerns.

When it comes to emerging technologies, there are several common concerns that often are Privacy is a major concern, as the collection and utilisation of personal data become more prevalent. The fear of job displacement due to automation is another significant concern, a technological advancements continue to reshape industries. Ethical concerns, such as the potential misuse of technology or the development of autonomous weapons, also loom large. Finally, there are concerns regarding the impact of emerging technologies on social inequality, as access and affordability may not be evenly distributed.

Unchecked hubris in the world of emerging technologies can have dire consequences. When we become overly confident in our abilities, we may neglect to thoroughly assess the potential risks and unintended consequences of our innovations. This can lead to the development and implementation of technologies without proper consideration for their impact on society, the environment, and individual well-being. In extreme cases, unchecked hubris can even result in catastrophic failures that could have been prevented with a more humble and cautious approach.

History is replete with examples of hubris in emerging technologies. One notable example is the development of nuclear power. In the early days, there was an unwavering belief in the limitless potential of nuclear energy. However, this hubris led to catastrophic accidents such as the Chernobyl and Fukushima disasters, which had severe consequences for both the environment and human health. Another example is the rise of social media platforms, when the initial focus on connectivity and free expression overshadowed concerns about the spread of misinformation and invasion of privacy. These examples serve as reminders of the dangers of unchecked hubris and the importance of recognising and addressing concerns.

To overcome hubris in the world of emerging technologies, it is crucial to recognise and address the concerns that accompany these advancements. One key approach is to prioritise ethical considerations throughout the entire development and implementation process. This involves actively assessing the potential risks and ethical implications of emerging technologies, as well as engaging in open and transparent discussions about these concerns. By acknowledging and actively seeking to address these concerns, we can ensure that emerging technologies are developed and utilised in a responsible and ethical manner.

Ethical considerations play a pivotal role in overcoming hubris in emerging technologies. essential to integrate ethical frameworks into the decision-making processes surrounding these advancements. This includes considering the potential consequences of emerging technologies on individuals, communities, and the environment. By adopting a more holistic and empathetic approach, we can mitigate the risks associated with unchecked hubris and strive towards more responsible and sustainable technological advancements.

Mitigating hubris requires a multi-faceted approach. One strategy is to foster collaboration and interdisciplinary approaches in the development and implementation of emerging technologies. By bringing together experts from various fields, we can tap into a diversity perspective and ensure a more comprehensive assessment of potential risks and considerations. Additionally, engaging in rigorous and independent peer review processes helps identify potential blind spots and ensure that technologies are thoroughly evaluated before being deployed.

Collaboration and interdisciplinary approaches are crucial in addressing concerns associated with emerging technologies. Oftentimes, the implications of these technologies extend beyond the realm of a single discipline. By working together, experts from different fields can identify and address potential risks, develop comprehensive guidelines, and promote responsible practices. This collaborative approach ensures that concerns are not overlooked or dismissed due to the hubris of a single discipline or individual.

Public engagement and transparency are vital in overcoming hubris and building trust in emerging technologies. By involving the public in the decision-making processes and openly communicating the potential risks and benefits, we can foster a sense of ownership and accountability. Transparency helps to dispel concerns and allows for a more informed and inclusive discussion about the development and implementation of emerging technologies. Furthermore, public engagement enables a diversity of perspectives to be considered, ensuring that the concerns of various stakeholders are taken into account.

Hubris in the world of emerging technologies pose significant challenges that must be recognised and addressed. By understanding the common concerns surrounding these technologies, we can actively work towards mitigating the dangers of unchecked hubris. Ethical considerations, collaboration, interdisciplinary approaches, and public engagement are key pillars in overcoming hubris. By embracing responsible and ethical practices, we can ensure that emerging technologies contribute to a more equitable, sustainable, and inclusive future. Let us remember that humility and a commitment to addressing concerns are essential in navigating the complex landscape of emerging technologies.

To navigate the promises and perils of emerging technologies, it is crucial to have robust regulations and policies in place. These regulations should strike a balance between fostering innovation and ensuring the responsible and ethical use of these technologies.

One approach is to establish clear guidelines and standards for the development and deployment of emerging technologies. This includes ethical frameworks, safety protocols, and guidelines for responsible conduct. For example, in the field of artificial intelligence, there is a growing call for transparency, accountability, and fairness in algorithmic decision-making. By setting clear standards, we can minimise the risks and maximise the benefits of these technologies.

Another approach is to promote interdisciplinary collaboration and dialogue among stakeholders. The development and deployment of emerging technologies require input from experts in various fields, including science, ethics, law, and policy. By fostering collaboration and dialogue, we can ensure that decisions are made with a comprehensive understanding of the potential risks and benefits.

Moreover, regulations and policies should be flexible and adaptable to keep pace with rapid technological advancements. Emerging technologies are evolving at an unprecedented rate and regulations must be able to respond to new challenges and emerging ethical concerns. This requires a proactive approach that anticipates potential risks and enables timely interventions.

To better understand the ethical considerations and debates surrounding playing God through emerging technologies, it is helpful to examine real-world examples. These case studies shed light on the potential benefits, risks, and ethical dilemmas associated with playing God.

One such case study is the use of gene editing technologies like CRISPR. This revolutionary tool allows scientists to modify the DNA of living organisms with unprecedented precision. While it holds the promise of curing genetic diseases and improving human health, it also raises ethical concerns about the potential misuse and unintended consequences. The case of the Chinese scientist who used CRISPR to edit the genes of human embryos without proper oversight sparked a global debate about the ethical boundaries of gene editing.

Another case study is the development of autonomous weapons systems powered by artificial intelligence. These weapons have the potential to make autonomous decisions about who to target and when to use lethal force. While they offer the promise of reducing human causalities and increasing military effectiveness, they also raise concerns about the loss of human con and the potential for indiscriminate harm. The development and deployment of autonomous weapons systems have sparked intense debates about the ethical implications of delegating life-and-death decisions to machines.

The concept of playing God through emerging technologies raises profound ethical considerations and sparks intense debates. These debates revolve around fundamental questions of human agency, the sanctity of life, and the potential consequences of our actions.

One of the key ethical considerations is the question of who gets to decide and control the technologies. The power to manipulate life and alter the course of evolution raises questions about democratic decision-making, accountability, and the distribution of benefits and risks. These technologies have the potential to exacerbate existing power imbalances and create forms of inequality if not carefully managed.

Another ethical consideration is the potential loss of human dignity and the erosion of what it means to be human. By playing God, we risk devaluing the uniqueness and inherent worth of being human. The pursuit of perfection and the desire to transcend our natural limitations come at the cost of our humanity. It is crucial to ensure that these technologies are developed and used in a way that respects and upholds human dignity.

Additionally, there is the question of long-term consequences and the potential for irreversible harm. The complex and interconnected nature of life and ecosystems means that our actions can have far-reaching and unpredictable impacts. As we gain the power to manipulate nature, it is essential to exercise caution and consider the potential consequences of our actions on future generations and the natural world.

The promises and perils of emerging technologies present us with a delicate balancing act between innovation and responsibility. On the one hand, these technologies offer immense opportunities for progress and improvement. On the other hand, they pose significant risks and ethical challenges that must be carefully managed.

To strike this balance, it is essential to adopt a proactive and responsible approach to innovation. This requires a comprehensive understanding of the potential risks and benefits of emerging technologies, as well as a commitment to ethical decision-making and responsible conduct. It also necessitates the involvement of diverse stakeholders, including scientists, ethicists, policymakers, and the general public.

Moreover, it is crucial to foster a culture of responsible innovation that values transparency, accountability, and inclusivity. This includes promoting open dialogue, encouraging interdisciplinary collaboration, and actively seeking input from diverse perspectives. By involving stakeholders in the decision-making process, we can ensure that emerging technologies are developed and used in a way that reflects societal values and addresses pressing global challenges.

To guide the evaluation and use of emerging technologies, ethical frameworks can provide valuable tools. These frameworks offer a systematic approach to ethical decision-making and help navigate the complex ethical considerations associated with playing God through emerging technologies.

One such framework is the principle of beneficence, which emphasises the importance of promoting the well-being and welfare of individuals and society. This principle requires careful consideration of the potential benefits and harms of emerging technologies and a commitment to maximising the overall good. It also calls for a proactive approach to identifying and mitigating potential risks and harms.

Another ethical framework is the principle of justice, which focusses on fairness and equity. This principle requires that the benefits and risks of emerging technologies are distributed in a just and equitable manner. It calls for equal access to these technologies, safeguards against discrimination, and the promotion of social justice. By applying the principle of justice, we can ensure that the potential benefits of emerging technologies are shared by all and that no one is left behind.

Moreover, the principle of autonomy plays a crucial role in ethical decision-making. This principle recognises the importance of individual autonomy and the right to make informed choices about one's own body and life. It calls for respect for individual decision-making and the protection of privacy and personal autonomy. By upholding the principle of autonomy can ensure that individuals have the freedom to make decisions about the use of emerging technologies that align with their values and beliefs.

The promises and perils of emerging technologies are intertwined with the ethical concern surrounding playing God. While these technologies hold immense promise for improving healthcare, enhancing communication, and addressing global challenges, they also raise profound ethical questions. The potential benefits of playing God through emerging technologies are vast, but so are the risks and dangers. It is crucial to navigate these challenges with a proactive and responsible approach, guided by robust regulations, interdisciplinary collaboration, and ethical frameworks. By striking a balance between innovation and responsibility, we can harness the power of emerging technologies for the betterment of humanity while ensuring that we do not overstep our moral boundaries.

https://orcid.org/0009-0005-0854-6213

CHAPTER 11

The Impact on Society

ARTIFICIAL INTELLIGENCE (AI) is increasingly an integral part of our lives, with its potential to revolutionise various industries. From enhancing cybersecurity to improving healthcare, AI offers a wide range of benefits. However, with great promise comes significant ethical challenges. Governments must consider the ethical dilemmas and moral questions associated with the deployment of AI and the impact it has on society.

One of the primary concerns surrounding AI is the potential displacement of workers by technology. Throughout history, automation has often replaced human labour in the short term but led to the creation of new jobs in the long run. However, there is widespread concern that AI and associated technologies could result in mass unemployment in the next two decades. A study suggests that new information technologies could put a substantial share of employment at risk across various occupations in the near future.

The impact of AI on the workforce is already evident in sectors such as finance, manufacturing, transportation, and healthcare. Unmanned vehicles and autonomous drones are performing tasks that previously required human intervention. While blue-collar jobs have already been affected by automation, advancements in AI could potentially impact a broader range of jobs. As computers become more sophisticated, creative, and versatile, more positions may become obsolete.

Economists are generally optimistic about the prospects of AI on economic growth. Robotics, for example, has contributed to annual GDP growth and labour productivity in several countries. However, the effect

 DOI: 10.1201/9781003502708-12

of AI and robotics on the workforce is still difficult to quantify as we are in the early stages of the technology revolution. Experts have conflicting views on the impact of emerging technologies. While some believe that robots and digital agents will displace significant numbers of workers, others expect that technology will create new jobs and industries.

The potential consequences of AI on the labour market are far-reaching. A study conducted by Frey and Osborne predicts that there is a high probability of 47% of US workers seeing their jobs become automated within the next two decades. Certain occupations, such as telemarketers, title examiners, and library technicians, have a higher likelihood of being computerised. On the other hand, occupations in fields like education, legal services, and arts are considered less amenable to automation.

AI and robotics advancements have the potential to streamline businesses, making them more efficient and productive. However, this efficiency may come at the expense of human workforces, leading to increased social inequalities. The ownership of AI-driven companies may result in disproportionate benefits for individuals, exacerbating existing economic divisions.

Changes in employment due to automation and digitisation will not only result in job losses but also affect job quality. New jobs created by AI may require high levels of skill but involve repetitive and monotonous tasks, resembling 'white-collar sweatshops'. The emergence of the gig economy has led to the exploitation of temporary workers who perform essential tasks in the AI value chain. These workers, known as 'mechanical Turks', engage in activities such as content moderation and data tagging. They often work outside the protection of labour laws and face precarious working conditions.

The equitable distribution of benefits and the impact of AI on workers' well-being are critical ethical issues. The involvement of temporary workers in the AI value chain is often hidden, and they are inadequately compensated for their contributions. These workers, both within and outside the EU and the United States, face challenges such as low wages, poor working conditions, and limited access to education and training. The discriminatory impact of AI on different demographics, including young people, minorities, and women, further exacerbates inequality.

With the increasing integration of AI into various aspects of society, concerns about privacy and data security have gained prominence. AI systems heavily rely on vast amounts of data for training and decision-making, raising questions about the collection, storage, and use of personal information. The unauthorised access or misuse of data can have severe consequences, including identity theft, fraud, and violation of privacy rights.

The use of AI in surveillance and monitoring also raises ethical concerns. Facial recognition technology, for example, can infringe upon individuals' privacy and civil liberties. The potential for mass surveillance and the abuse of AI-powered systems by governments or other entities is a significant challenge that needs to be addressed.

To ensure the ethical use of AI, it is crucial to establish robust data protection regulations and frameworks. Transparency and accountability should be prioritised, allowing individuals to have control over their personal information. Companies and organisations must implement stringent security measures to safeguard data and prevent unauthorised access. Additionally, policymakers should develop regulations that balance the benefits of AI with privacy rights and civil liberties.

Another ethical issue related to AI is the potential for bias and discrimination in decision-making algorithms. AI systems learn from vast datasets, which may contain inherent biases and reflect societal prejudices. If these biases are not identified and addressed, AI can perpetuate discrimination in various domains, including hiring, lending, and criminal justice.

To mitigate bias and discrimination, AI developers and researchers must ensure that datasets used for training are diverse, representative, and free from bias. Ethical guidelines and standards should be established to promote fairness and non-discrimination in AI systems.

Regular audits and evaluations of AI algorithms can help identify and rectify biases.

Moreover, it is essential to have diverse and inclusive teams involved in the development and deployment of AI systems. By incorporating different perspectives and experiences, biases can be minimised, and the potential for discrimination reduced.

As AI systems become more complex and autonomous, the lack of accountability and transparency poses significant ethical challenges. When AI makes decisions that impact individuals' lives, it is crucial to understand how those decisions are made and hold responsible parties accountable.

Currently, many AI algorithms operate as "black boxes," with their decision-making processes hidden and often poorly understood. This lack of transparency makes it difficult to trace errors or biases and can lead to unjust outcomes. To ensure ethical AI deployment, it is essential to develop explainable and interpretable AI systems. The ability to understand and explain how AI arrives at its decisions is vital for building trust and ensuring accountability.

Furthermore, there should be clear guidelines and regulations regarding the responsibility of individuals and organisations when AI systems make mistakes or cause harm. Establishing legal frameworks and ethical standards can help address issues of accountability and ensure that AI is used in a responsible and transparent manner.

The widespread adoption of AI has significant implications for trust and human autonomy. As AI systems become more integrated into our daily lives, individuals may rely on them for critical decision-making, such as medical diagnoses or financial advice. However, blind trust in AI systems can erode human agency and decision-making capabilities.

Maintaining human autonomy in the age of AI requires striking a balance between human judgement and reliance on AI systems. It is crucial to ensure that individuals have access to understandable explanations and have the ability to question, challenge, or override AI decisions when necessary. Building trust in AI systems requires transparent communication about their limitations and potential biases, allowing individuals to make informed decisions.

Additionally, safeguards should be in place to prevent the misuse or abuse of AI systems that could potentially manipulate individuals' thoughts, emotions, or behaviours. The ethical considerations surrounding AI and its impact on trust and human autonomy should be at the forefront of development and implementation strategies.

While AI offers numerous benefits, its deployment can also have significant environmental consequences. The increasing reliance on AI technologies demands vast amounts of energy, contributing to higher carbon emissions and environmental degradation.

To mitigate the environmental impact of AI, it is essential to develop energy-efficient algorithms and hardware. Investing in renewable energy sources and optimising data centres can help reduce the carbon footprint associated with AI technologies. Furthermore, the responsible disposal and recycling of outdated AI equipment can minimise electronic waste and promote sustainability.

The ethical dilemmas and moral questions associated with the deployment of AI highlight the need for responsible and ethical practices. As AI continues to advance, it is crucial to address issues such as job displacement, inequality, privacy, bias, accountability, trust, human autonomy, and environmental impact. By incorporating ethical considerations into AI development and implementation, we can ensure that AI technology benefits society while avoiding or mitigating potential harm. The responsible and transparent use of AI is essential for creating an inclusive and equitable future.

https://orcid.org/0009-0005-0854-6213

CHAPTER 12

The Power of AI in Financial Services

ARTIFICIAL INTELLIGENCE (AI) has emerged as a transformative technology WITH the potential to revolutionise various industries, including financial services. The financial services industry plays a crucial role in the economy, affecting every citizen's life and contributing significantly to economic output. As AI continues to evolve, it is essential to assess its impact on the financial services sector and address the potential risks associated with its adoption.

Impact of AI in financial services: AI has the potential to bring numerous benefits to consumers, firms, and the financial system as a whole. For consumers, AI can enable more personalised financial products and services, enhancing the customer experience. Firms can leverage AI to improve predictive capabilities, streamline processes, and increase profitability. From a systemic perspective, AI can contribute to better risk management, fraud detection, and more efficient compliance with regulatory requirements.

Current and potential use of AI in financial services: The financial services industry has already begun adopting AI technologies, with machine learning (ML) applications being utilised by numerous firms. ML algorithms enable automation, data analysis, and decision-making, enhancing efficiency and accuracy. In the wholesale and institutional

DOI: 10.1201/9781003502708-13

markets, AI can optimise trading strategies, risk assessment, and regulatory compliance. Retail financial services, on the other hand, benefit from AI-driven automation in areas such as credit assessment, underwriting, and claims processing.

The end-user perspective: AI's impact on end-users is primarily felt through customer-facing applications such as chatbots and personalised financial products. AI-powered natural language processing and voice recognition technologies enable seamless customer interactions and improved accessibility. Additionally, AI can assist in the transition to a net-zero economy by analysing climate risks and optimising processes in sectors such as energy and insurance.

Potential risks from AI in financial services: While AI presents significant opportunities, it also poses unique risks that need to be carefully managed. These risks encompass areas such as protecting the integrity of financial markets, systemic risks, cybersecurity threats, misinformation, consumer protection, financial exclusion, and environmental concerns.

Protecting the integrity of financial markets: Ensuring the integrity of financial markets is a key objective for regulators. The adoption of AI in financial services introduces potential conflicts with this objective. For instance, if AI-driven trading algorithms are based on inaccurate data, it can lead to misallocation of resources and undermine market stability. The opacity of advanced AI models can also raise concerns about herd mentality bias and mispricing risk, eroding public trust in the financial system. This poses a risk to the overall integrity of financial markets and their ability to support the economy.

Systemic risks: The complexity and interconnectedness of financial markets make them susceptible to systemic risks arising from AI adoption. Instability and the potential erosion of public trust can disrupt market functioning and increase the likelihood of financial crises.

Cybersecurity risks also pose a significant systemic threat, with AI-based systems being vulnerable to data poisoning attacks and cybercriminal activities. The spread of misinformation through AI-generated content further exacerbates systemic risks.

Cybersecurity risks: The growing reliance on AI technologies introduces new avenues for cyberattacks in the financial services sector. Attackers can exploit vulnerabilities in AI systems, manipulate data inputs, and compromise decision-making processes, leading to harmful outcomes. Financial firms must maintain constant oversight of AI algorithms to mitigate cybersecurity risks effectively.

Misinformation: Data-driven AI models are susceptible to biases and inaccuracies if trained on corrupt or misleading data. In financial services, this can lead to poor decision-making and unethical outcomes. The spread of misinformation, particularly through social media, can impact price formation in global markets and undermine market integrity. Regulators need to address the risks associated with AI-generated misinformation and prevent bad actors from manipulating markets.

Consumer protection: Consumer protection is a crucial aspect of financial regulation, and the adoption of AI introduces new challenges in this area. The complexity and opacity of AI decision-making algorithms make it difficult for regulators to assess fairness, transparency, and compliance with consumer protection regulations. Biases in AI models can result in discriminatory outcomes, affecting access to financial products and services for certain groups. Privacy concerns also arise from the extensive use of personal data in AI applications, requiring robust data protection measures.

Financial exclusion: While AI can enhance financial inclusion by providing affordable and personalised products, it can also exacerbate existing inequalities. If AI algorithms are not designed to address biases and ensure fair access, certain groups may face exclusion from financial services. It is crucial to evaluate the impact of AI on financial inclusion and implement measures to mitigate exclusionary risks.

Environmental concerns: AI can contribute to the transition to a net-zero economy by optimising processes and analysing climate risks. However, the energy consumption associated with AI infrastructure and data processing raises environmental concerns. Regulators need to balance the potential benefits of AI with its environmental impact and encourage sustainable AI practices in the financial services industry.

Regulating AI in financial services: Given the potential risks associated with AI in financial services, effective regulation is crucial to ensure responsible and ethical AI adoption. A precautionary regulatory approach that anticipates potential risks and a risk-based framework are necessary to prevent harm. Collaboration between government, regulators, civil society organisations, and the public is essential to ensure fair and inclusive AI regulation. Engaging diverse perspectives will lead to better policy outcomes and protect against the influence of powerful tech and financial services industries.

AI is set to transform the financial services industry, offering significant benefits but also introducing unique risks. It is vital to manage these risks effectively through robust regulation, transparency, and ethical practices. Regulators must address the challenges associated with AI adoption, protect consumers, uphold market integrity, and ensure a fair and inclusive financial system. By striking the right balance between innovation and risk management, the financial services industry can leverage the full potential of AI while safeguarding the public interest and promoting sustainable growth.

Technology has transformative power that has touched every aspect of human existence, reshaping industries and creating new possibilities. Society must explore how technology is changing industries and examine the ethical implications of these advancements. It is incumbent on us to analyse the impact of technology on power dynamics, how it could transform industries, discuss the future potential of technology, and ultimately, consider the importance of balancing the benefits and risks while harnessing its power for positive change.

Industries across the globe are being revolutionised by technology, leading to enhanced efficiency, productivity, and innovation. From manufacturing to healthcare, retail to transportation, the integration of technology has disrupted traditional practices and opened up new opportunities. Automation and AI have streamlined processes, reducing human error and increasing accuracy, while data analytics and ML have unlocked valuable insights and predictive capabilities. As a result, companies are able to optimise their operations, deliver products and services more effectively, and ultimately, stay ahead in an increasingly competitive landscape.

However, the transformation brought about by technology is not without its challenges. While efficiency and convenience are undeniable benefits, there are ethical considerations that must be addressed.

As technology continues to advance at an unprecedented pace, it raises important ethical questions that demand our attention. One of the key concerns is the impact on employment. Automation and AI have the potential to replace human workers, leading to job displacement and economic inequality. It is crucial to ensure that as industries embrace technology, there are measures in place to retrain and upskill the workforce, ensuring that no one is left behind.

Another ethical consideration is privacy and data security. With the increasing reliance on technology, personal data is being collected and analysed on a massive scale. This raises concerns about data breaches, surveillance, and the potential for misuse. Safeguarding personal information and establishing robust data protection regulations are essential to maintain trust and ensure the responsible use of technology.

Furthermore, the rise of powerful technology companies has sparked debates around monopolies and the concentration of power. These companies have access to vast amounts of data, enabling them to influence decision-making processes and control markets. Striking a balance between innovation and fair competition is crucial to prevent the abuse of power and maintain a healthy, diverse business ecosystem.

Technology has the ability to reshape power dynamics within industries. Traditionally, power was concentrated in the hands of a few, often limited to large corporations or government entities. However, with the democratisation of technology, smaller players now have the opportunity to disrupt established industries and challenge the status quo. Start-ups and entrepreneurs can leverage technological advancements to create innovative solutions, levelling the playing field and promoting competition.

Additionally, technology has empowered individuals by providing them with access to information and platforms to voice their opinions. Social media, for instance, has become a powerful tool for activism and mobilisation, allowing citizens to hold corporations and governments accountable. This shift in power dynamics has the potential to drive positive change and foster a more inclusive and participatory society.

Technology even has the potential to change the dynamics of entire societies. In WWII, the entire course of the conflict and indeed the future of many countries across the world was determined by technological advancements.

Numerous industries have already experienced significant transformations due to technology. Therefore, we can explore a few examples:

Healthcare: The healthcare industry has been revolutionised by technology, leading to improved patient care, enhanced diagnostics, and increased accessibility. Electronic health records have streamlined medical documentation, enabling healthcare professionals to access patient information quickly and securely. Telemedicine has made healthcare more accessible, particularly for remote areas and underserved communities. Furthermore, advancements in medical imaging and genomics have revolutionised diagnostics and personalised medicine, allowing for more precise and targeted treatments.

Transportation: The transportation industry has undergone a massive transformation with the advent of technology. Ride-sharing platforms have disrupted the traditional taxi industry, providing convenient and cost-effective transportation options. Electric and autonomous vehicles are revolutionising the way we travel, reducing carbon emissions and increasing safety. Additionally, logistics and supply chain management have been optimised through the use of data analytics and real-time tracking, improving efficiency and reducing costs.

Retail: E-commerce has completely transformed the retail industry, offering consumers the convenience of shopping from anywhere, at any time. Online marketplaces have provided small businesses with a global reach, enabling them to compete with established retailers. Personalisation and recommendation algorithms have enhanced the shopping experience, tailoring product suggestions to individual preferences. Furthermore, technologies such as augmented reality have allowed customers to virtually try on products before making a purchase, bridging the gap between online and offline shopping.

The future of technology holds immense potential for further transformation. As advancements in AI, robotics, and quantum computing continue to accelerate, we can expect even greater disruption across industries. The Internet of Things (IoT) will connect everything from household appliances to industrial machinery, creating new opportunities for automation and data-driven decision-making. Virtual reality and augmented reality will revolutionise entertainment, education, and even remote work.

Furthermore, breakthroughs in clean energy and sustainable technologies will help address pressing environmental challenges.

However, it is important to approach these advancements with caution and consider their impact on society as a whole.

While technology offers immense benefits, it also carries risks that must be managed. As we embrace the transformative power of technology, it is crucial to strike a balance between innovation and responsibility. This requires a comprehensive approach that considers both short-term and long-term implications.

Regulatory frameworks need to keep pace with technological advancements to ensure accountability and protect consumer rights. Collaboration between governments, industry experts, and academia is essential to develop robust policies that address ethical concerns, promote fair competition, and safeguard privacy and data security.

Education and upskilling are also critical to equip individuals with the necessary knowledge and skills to thrive in a technology-driven world. Lifelong learning programmes and initiatives can help bridge the digital divide and ensure that everyone has equal opportunities to benefit from technological advancements.

Technology has the potential to be a powerful force for positive change. By leveraging its capabilities, we can address some of the most pressing global challenges, such as poverty, inequality, and climate change. However, to achieve this, ethical considerations must be at the forefront of technological advancements.

The role of ethics in guiding technological development cannot be overstated. It is essential to ensure that technology is designed and implemented with the best interests of humanity in mind. Ethical frameworks need to be established to guide decision-making processes and prevent the misuse of technology. Transparency and accountability should be embedded into the development and deployment of new technologies, allowing for public scrutiny and input.

Technology has the power to transform industries, reshape power dynamics, and create new possibilities. However, it is essential to approach these advancements with caution and address the ethical implications they bring. By balancing the benefits and risks of technology, we can harness its power for positive change and create a future that is inclusive, sustainable, and equitable. Let us embrace the transformative power of technology and work together to shape a better world.

https://orcid.org/0009-0005-0854-6213

CHAPTER 13

Cybersecurity Considerations with AI

IN THE MODERN ERA of cybersecurity, the landscape of cyberattacks has undergone a significant transformation. While manual efforts and traditional methods were once the primary tools of threat actors, the integration of artificial intelligence (AI) has taken cyber threats to a whole new level. The marriage of AI and cyber threats has empowered even those with basic technical skills to execute successful attacks, leading to an alarming increase in the number and sophistication of cyberattacks.

According to an IBM report, the average cost of a data breach reached a record high of USD 4.45 million in 2023,[1] and with the growing involvement of AI in cyberattacks, these numbers are expected to continue rising. The expansion of remote work due to the pandemic has further widened the attack surface, making organisations vulnerable to breaches from both well-known threat groups and lesser-known adversaries. AI has become a key tool at every stage of a cyberattack, from reconnaissance to exfiltration.

Technology leaders must explore the methods and risks associated with AI-driven attacks in the evolving cyberspace. We will delve into the various stages of AI-powered attack scenarios, analyse the potential threats AI-driven attacks hold for the future, discuss the economic implications of the growing cyberattack space, and finally, outline strategies to combat AI-driven attacks.

DOI: 10.1201/9781003502708-14

Technology leaders must explore the methods and risks associated with AI-driven attacks in the evolving cyberspace. We will delve into the various stages of AI-powered attack scenarios, analyse the potential threats AI-driven attacks hold for the future, discuss the economic implications of the growing cyberattack space, and finally, outline strategies to combat AI-driven attacks.

Cyberattacks have continuously evolved from manual to AI-powered threats. To understand the exponential growth of the cyberspace and the associated threats, we must first take a trip back in time to the birth of cyberattacks. In the 1980s, cyberattacks emerged as a frequent and alarming threat with the introduction of worms and viruses like the Trojan Horse. This marked the beginning of the antivirus software era, as the war against cyber threats escalated.

The 1990s witnessed the advent of the internet, which opened doors to a plethora of cyber threats. Polymorphic viruses, capable of mutating as they spread through computing systems, became a significant concern. To combat these evolving threats, new methods of securing communications were devised, leading to the development of secure sockets layer (SSL) to encrypt data between parties and ensure secure internet connections.

The Internet era brought not only many benefits but also risks like modern malware. As the 21st century dawned, the widespread availability of reliable broadband led to a surge in internet usage globally. However, this also resulted in a corresponding increase in vulnerabilities and infections in the cyberspace. Modern malware, such as email-based infections and social engineering attacks, began to emerge. Hacking credit card information became a major focus for cybercriminals, prompting companies to develop defensive cybersecurity measures, including open-source antivirus software.

The 2010s witnessed a significant shift in the balance of power, with adversaries surpassing cybersecurity efforts and causing substantial financial losses for businesses and governments. High-profile breaches, such as the global payment systems data breach in 2012, the Yahoo data breach in 2013–14, and the WannaCry ransomware attack in 2017, highlighted the vulnerability of organisations. During this time, vendors developed innovative approaches like multi-factor authentication and network behavioural analysis to detect behavioural anomalies in files.

AI and machine learning (ML) will play a crucial role in changing the cyber landscape. While AI and ML have been present in the cybersecurity field since the 1950s, their involvement in cyberattacks was initially underestimated. However, as AI and ML technologies have advanced, their significance in the cybersecurity space has become increasingly apparent. AI has enabled the automation of malicious activities, the tailoring of attack strategies, and the exploitation of vulnerabilities with greater efficiency.

Generative AI, in particular, has revolutionised the cyber threat landscape. Adversaries can now use text-based generative AI to explore endless possibilities for attack methods and automate models to evade defences. This has led to notable AI-powered cyberattacks, such as the TaskRabbit cybersecurity breach, the Nokia breach, and the WordPress data breach. In response, the cybersecurity industry has been actively developing innovative defence strategies, including security information and event management (SIEM) solutions, to protect organisational networks.

Security implications of AI-powered attacks are significant. The integration of AI into cyberattacks brings about unique advantages that traditional manual methods cannot replicate. We can explore some of the security implications of AI-powered attack scenarios.

Automation and high scaling: One of the key advantages of AI-powered attacks is the ability to automate various stages of the attack process, including reconnaissance, vulnerability scanning, and exploitation. Automation allows attackers to target multiple systems simultaneously, making AI-powered attacks highly scalable. Traditional attacks, on the other hand, require more manual effort and are limited in terms of scale.

Efficiency and adaptability: AI systems can analyse vast amounts of data in real-time, enabling attackers to identify vulnerabilities rapidly and adapt their attack strategies accordingly. Tools like PassGAN, which uses ML to generate password guesses, can carry out attacks at a rapid pace. AI algorithms can also evade defence mechanisms by using vast training data and adapting to countermeasures.

Highly sophisticated malware: AI-powered attacks can employ sophisticated evasion techniques to bypass traditional security measures. Modern malware is capable of bypassing server filters and continuously mutating to evade analysis by defenders. Notable examples include IBM's DeepLocker, a proof-of-concept malware variant that leverages ML algorithms for cyberattacks.

Malware-infected botnets: Malware is no longer limited to infecting individual systems. The rise of botnets, networks of malware-infected devices, has become increasingly prevalent.

These botnets can scan the entire internet for vulnerabilities, targeting multiple sites simultaneously. AI algorithms optimise the command-and-control infrastructure of these botnets, making the malware more resilient and difficult to trace.

Generative AIs in the picture: AI systems can analyse personal information and social media profiles to generate tailored phishing emails and voice messages that appear legitimate.

Generative AI can scour the internet for e-books, articles, websites, and posts, gathering information that can be used to target and profile victims. Unfortunately, generative AI can also be exploited to write malware pseudo-code, posing a significant threat.

AI-powered attack scenarios can be broken down. To fully understand the impact of AI-driven attacks, it is essential to examine the various stages of an attack. We can delve into each stage and explore the role of AI in enabling adversaries to execute successful attacks.

Reconnaissance: In the reconnaissance stage, attackers rely on AI to automate and enhance the process of profiling targets, scanning for vulnerabilities, and framing the entire attack. AI's pattern recognition capabilities enable the identification of links and correlations that may elude human analysts. By analysing publicly available information from sources such as social media, forums, and leaked databases, AI-powered bots can quickly gather valuable data.

Initial access: The initial access stage is crucial for attackers to gain a foothold in an organisation's network. AI plays a significant role in this stage, particularly in social engineering exploitation methods and generating effective synthetic phishing URLs. AI algorithms can analyse vast datasets of leaked passwords and user behaviours to crack passwords more efficiently.

Privilege escalation: Privilege escalation involves obtaining higher-level access and privileges within a target network. AI can identify user patterns indicating privileged accounts or high-level access.

AI-powered tools can scan networks for access control vulnerabilities, providing attackers with instant access to potential weak spots. Deep-reinforcement learning can automate privilege escalation.

Persistence and lateral movement: During the persistence and lateral movement stage, adversaries aim to expand their presence within a compromised network while maintaining a strong foothold. AI models, such as generative adversarial networks, can generate undetectable adversarial malware to bypass black-box detection systems. AI-powered tools can scan and map networks, identifying connected devices, services, and vulnerabilities.

Command and control: The command-and-control stage is where adversaries establish communication and control over compromised systems without leaving traces. AI algorithms can obfuscate communication channels by generating malicious traffic or behaviour that mimics legitimate patterns. AI also enables attackers to automate responses and adapt strategies in real-time, making detection and tracing more challenging.

Exfiltration: Exfiltration involves stealing confidential data from an organisation. Adversarial AI eliminates the need for extensive knowledge of data exfiltration techniques, as AI can assist in generating exfiltration codes. AI can also optimise the exfiltration traffic, split it across multiple channels, or utilise covert communication channels to mask data transfers.

The future of AI-driven attacks means there are many more potential threats on the horizon. AI-driven attacks are continuously evolving, and threat actors are constantly innovating and adapting their tactics to breach digital defences. We can explore some emerging attack methods that exploit the transformative potential of AI in the cyber landscape.

Search engine optimisation and malvertising: Threat actors leverage search engine optimisation (SEO) techniques and paid advertising methods to improve page rankings and increase the exposure of their fake pages to the public. When users click on these links and download files, malicious JavaScript is loaded into their devices, granting control to attackers. This technique is known as malvertising and is classified as a new MITRE ATT&CK technique.

Scaling up attacks with adversarial AI: Adversarial AI not only enables attackers to conduct large-scale attacks with relative ease but also amplifies the impact of their malicious activities. Adversaries can create new malware attack scripts effortlessly, and the discovery of zero-day vulnerabilities in complex environments requires minimal effort. As a result, organisations face an increased risk of attacks.

Social engineering attacks using AI: The heightened risk of social engineering attacks is a major concern in the cybersecurity landscape. Adversaries can analyse vast datasets and create highly customised and persuasive phishing emails, making it difficult for users to differentiate legitimate communications from fraudulent ones. The multilingual capability of generative AI allows adversaries from different regions to engage in social engineering attacks, expanding the pool of potential threats.

Economic implications of the growing cyberattack space: The growing cyberattack space carries significant economic implications for organisations. With the ongoing economic crisis and budget cuts in IT departments, adversaries are capitalising on the integration and development of AI in cyberattacks, intensifying the financial impact of cybercrime. The cost of remediation, recovery, and regulatory compliance following a cyberattack has increased, and organisations also face hefty non-compliance penalties due to regional mandates. Mitigating financial losses and protecting organisational stability require strategic measures.

Combating AI-driven attacks means we must develop strategies for detection, mitigation, and compliance. To counter the exponential growth of AI-driven attacks, organisations must adopt a proactive and adaptive approach to cybersecurity. We can explore strategies for detecting, mitigating, and complying with AI-driven attacks.

Detection stage: In the detection stage, organisations should leverage comprehensive SIEM solutions. These solutions employ advanced analytics, anomaly detection, and behavioural analysis to detect indicators of compromise (IoCs) at each stage of an AI-powered attack. Real-time alert engines and correlation engines help security teams maintain constant vigilance and promptly respond to threats.

Mitigation stage: Mitigating AI-driven attacks requires continuous monitoring, advanced threat intelligence, and an incident response framework. SIEM solutions like ManageEngine Log360 provide incident response features that automatically initiate predefined workflows to counteract threats. Instant notifications and ticketing systems ensure that security personnel are immediately aware of concerning activities and can take prompt action.

Compliance and health check: To meet compliance requirements, organisations can rely on SIEM solutions, which offer audit-ready report templates for a wide range of policies. These solutions simplify compliance reporting by providing intuitive dashboards that display metrics demonstrating compliance with standards such as PCI DSS, HIPAA, and GDPR. Custom compliance reports can address both external mandates and internal compliance needs effectively.

AI-driven attacks have ushered in a new era of sophisticated and relentless cyber threats. Adversarial AI has become a powerful tool for cyber adversaries, automating and scaling their malicious activities with unprecedented efficiency. To navigate this evolving landscape, organisations must adopt proactive and adaptive strategies, leveraging advanced security solutions like SIEM to detect, mitigate, and comply with AI-driven attacks. By combining advanced technology with robust incident management and risk security postures, defenders can anticipate and counter emerging adversarial AI threats, creating a safer digital realm for generations to come.

AI systems have the potential to bring numerous benefits to society. However, to fully realise the opportunities of AI, it is crucial to develop, deploy, and operate these systems in a secure and responsible manner. Cybersecurity plays a vital role in ensuring the safety, resilience, privacy, fairness, efficacy, and reliability of AI systems. We can explore the guidelines for developing secure AI systems that function as intended, protect sensitive data, and are available when needed.

AI systems are subject to unique security vulnerabilities that must be considered alongside standard cybersecurity threats. The rapid pace of AI development often prioritises functionality over security, making it imperative to incorporate security as a core requirement throughout the system's lifecycle. The guidelines outlined are based on industry best practices and are designed to mitigate risks and enhance the security of AI systems.

SECURE DESIGN

Secure design is a critical aspect of developing AI systems. By understanding risks, conducting threat modelling, and making informed design choices, developers can create robust and secure AI systems. Some key considerations in secure design include:

1. *Staff awareness*: It is essential for system owners, senior leaders, data scientists, and developers to understand the threats and risks associated with secure AI. Regular training, awareness programmes, and adherence to secure coding techniques contribute to a security-focused mindset.
2. *Threat modelling*: Assessing potential threats and their impacts is crucial in managing risk.

   A holistic approach should encompass AI-specific threats, such as adversarial ML, along with standard cybersecurity threats.
3. *Security in design choices*: When designing AI systems, factors like supply chain security, model architecture, choice of models, and integration of external components should be considered. Supply chain security includes evaluating the security posture of suppliers and ensuring the integrity of components.
4. *Security benefits and trade-offs*: The selection of AI models involves a balance between various requirements. Factors like model complexity, appropriateness for the use case, interpretability, and supply chain provenance should be considered. Privacy-enhancing techniques and model hardening can also enhance security.

SECURE DEVELOPMENT

The development stage of AI system development requires attention to supply chain security, asset management, and mitigation of technical debt. Key guidelines for secure development include:

1. *Secure supply chain*: Assessing and monitoring the security of AI supply chains is crucial. Suppliers should adhere to the same security standards applied to other software components. Well-secured and documented hardware and software components should be acquired from trusted sources.

2. *Asset management*: Recognising the value of AI-related assets, such as models, data, documentation, and logs, enables effective asset tracking and protection. Processes and tools for version control, authentication, and secure storage should be implemented.

3. *Technical debt management*: Managing technical debt is essential throughout the AI system's lifecycle. Proactive identification, tracking, and mitigation of technical debt contribute to long-term benefits and reduce security risks.

SECURE DEPLOYMENT

The deployment stage involves protecting infrastructure, models, and ensuring responsible release. Key guidelines for secure deployment include:

1. *Infrastructure security*: Applying good infrastructure security principles is vital at every stage of the AI system's lifecycle. Access controls, segregation of environments, and protection against standard cyberattacks safeguard the system's integrity.

2. *Model protection*: Protecting AI models from direct and indirect access is crucial.

   Implementing standard cybersecurity measures and controlling the query interface help prevent unauthorised access, modification, and exfiltration of confidential information.

3. *Incident management*: Developing incident response plans and escalation procedures is essential to address security incidents promptly. Training responders, providing high-quality audit logs, and enabling incident response processes contribute to effective incident management.

4. *Responsible release*: Models, applications, and systems should undergo appropriate security evaluations, benchmarking, and red teaming before release. Transparency about known limitations or potential failure modes helps users make informed decisions.

SECURE OPERATION AND MAINTENANCE

In the operation and maintenance stage, monitoring system behaviour, managing updates, and sharing lessons learnt are critical. Key guidelines for secure operation and maintenance include:

1. *System monitoring*: Monitoring the outputs and performance of AI models and systems helps identify potential intrusions, compromises, and data drift. It enables prompt responses to security incidents and ensures system integrity.

2. *Input monitoring*: Monitoring and logging system inputs, such as inference requests and queries, aids compliance, audit, and investigation in case of compromise or misuse. Detection of out-of-distribution and adversarial inputs enhances system security.

3. *Secure updates*: Implementing automated updates by default and using secure update procedures help distribute updates effectively. Changes to data, models, or prompts should be treated as potential system behaviour changes, and users should be supported in evaluating and responding to these changes.

4. *Information sharing*: Participating in information-sharing communities and maintaining open lines of communication for feedback helps improve system security. Collaborating with the wider community, responding to vulnerability disclosures, and taking prompt mitigation actions contribute to a secure AI ecosystem.

Implementing these guidelines for secure AI system development helps mitigate risks, enhance system security, and promote responsible use of AI. By adopting a secure-by-design approach and adhering to industry best practices, organisations can build AI systems that are robust, reliable, and protect sensitive information.

Developing secure AI systems requires a holistic approach that considers the unique security vulnerabilities of AI. By following the guidelines for secure design, development, deployment, and operation, organisations can mitigate risks and ensure the integrity, availability, and confidentiality of AI systems. Embracing a security-first mindset, fostering staff awareness, and implementing robust security measures contribute to the responsible development and use of AI.

AI and ML technologies are rapidly advancing, revolutionising various industries and offering countless beneficial applications. However, as these technologies continue to evolve, it is crucial to acknowledge the potential security threats that may arise from their malicious use. We must explore the landscape of these threats, propose ways to mitigate them, and highlight the importance of collaboration between policymakers, researchers, and experts in the field of AI.

As AI capabilities become more powerful and accessible, we can expect both the expansion of existing threats and the emergence of new ones. The scalable nature of AI systems enables malicious actors to carry out attacks more efficiently and at a larger scale, broadening the range of potential targets. Additionally, AI can be leveraged to complete tasks that would be impractical for humans, opening up new avenues for attacks. Furthermore, the effectiveness of attacks enabled by AI is expected to increase, as they become finely targeted, difficult to attribute, and exploit vulnerabilities in AI systems.

To fully grasp the potential threats, it is essential to analyse various security domains and understand the changes that AI can bring to each.

Digital security: In the realm of digital security, AI can be employed to automate tasks involved in cyberattacks. This could significantly lower the costs associated with labour-intensive attacks, such as spear phishing. Moreover, AI systems can exploit human vulnerabilities, software vulnerabilities, and even the vulnerabilities of other AI systems. For example, speech synthesis can be used for impersonation, automated hacking can exploit software vulnerabilities, and adversarial examples and data poisoning can undermine AI systems.

Physical security: AI's impact on physical security is notable, with the automation of tasks involved in attacks using drones and other physical systems. Autonomous weapons systems, enabled by AI, can expand the threats associated with such attacks. Furthermore, novel attacks can be devised that subvert cyber-physical systems or involve physical systems that are difficult to control remotely. An example of this would be a swarm of thousands of micro-drones used for malicious purposes.

Political security: The use of AI to automate tasks related to surveillance, persuasion, and deception introduces new threats to political security. Mass-collected data can be analysed using AI, compromising privacy. Targeted propaganda can be created to manipulate public opinion, and the manipulation of videos can deceive individuals. These threats are particularly concerning in authoritarian states, but they can also undermine the ability of democracies to sustain truthful public debates.

To address the evolving threat landscape, it is crucial for policymakers, researchers, and AI practitioners to collaborate closely. By working together, they can investigate, prevent, and mitigate potential malicious uses of AI. Here are four high-level recommendations to guide their efforts:

1. *Collaborative efforts*: Policymakers should actively collaborate with technical researchers to understand and address the potential security risks associated with AI. This collaboration can facilitate the development of effective preventive measures.
2. *Dual-use considerations*: Researchers and engineers in AI should take the dual-use nature of their work seriously. They must prioritise the prevention of harmful applications and proactively engage with relevant actors when foreseeing potential risks.
3. *Importing best practices*: Best practices from other domains, such as computer security, should be identified and imported to address dual-use concerns in AI. By leveraging existing frameworks and methodologies, AI practitioners can enhance the security of AI systems.
4. *Diverse stakeholder engagement*: In discussions surrounding the challenges posed by malicious AI use, it is crucial to involve a wide range of stakeholders and domain experts. Their diverse perspectives can contribute to the development of comprehensive solutions.

In addition to these high-level recommendations, there are several priority research areas that require further exploration and intervention. These research areas include learning from and collaborating with the cybersecurity community, exploring different openness models for AI research, promoting a culture of responsibility among AI researchers and organisations, and developing technological and policy solutions to build a safer future with AI.

To effectively combat the potential threats posed by AI, it is essential to learn from and collaborate with the cybersecurity community. By embracing practices such as red teaming, formal verification, responsible disclosure of AI vulnerabilities, and the use of security tools and secure hardware, the AI community can enhance the resilience of AI systems against malicious attacks. This collaboration can help identify vulnerabilities, develop effective countermeasures, and ensure the security of AI technologies.

Exploring different openness models is crucial. As the dual-use nature of AI becomes more apparent, it is crucial to reimagine norms and institutions surrounding the openness of research. Pre-publication risk assessment in technical areas of special concern can help identify potential risks early on. Central access licensing models and sharing regimes can prioritise safety and security. Lessons from other dual-use technologies can guide the development of frameworks that promote responsible and secure AI research.

Promoting a culture of responsibility is essential. AI researchers and the organisations that employ them have a significant role in shaping the security landscape of the AI-enabled world. Emphasising education, ethical standards, and expectations, AI practitioners can foster a culture of responsibility. By proactively addressing the potential security risks associated with AI, researchers and organisations can contribute to a more secure and trustworthy AI ecosystem.

In addition to collaborative efforts and responsible practices, it is crucial to explore technological and policy solutions to mitigate the malicious use of AI. Research in privacy protection, coordinated use of AI for public-good security, monitoring AI-relevant resources, and legislative and regulatory responses can all contribute to a safer future with AI. By leveraging these solutions, policymakers can establish a robust framework that ensures the responsible and secure use of AI technologies.

AI and ML have already been widely adopted in various applications, such as automatic speech recognition, machine translation, and search engines. However, the increasing development of AI also raises concerns about its potential malicious use. We must explore the interaction of AI and cybersecurity, analysing the risks and potential threats associated with the malicious use of AI.

The malicious use of AI can pose threats to digital, physical, and political security. Criminals can exploit AI technologies to enhance their hacking and social engineering techniques, leading to increased risks of cyberattacks. Non-state actors may weaponise consumer drones, compromising physical security. Additionally, AI-enabled surveillance, profiling, and disinformation campaigns can undermine political security.

To address these risks, it is important to understand how AI can be used maliciously and develop strategies to prevent and mitigate the potential harmful effects.

This book focusses on AI technologies that are currently available or plausible in the next five years. It primarily considers scenarios where individuals or organisations deploy AI technology or compromise AI systems to undermine the security of others.

While the threat of malicious use of AI has been highlighted in various settings, the comprehensive analysis of AI and malicious intent is still limited. Previous studies have focused more on unintentional forms of AI misuse, such as algorithmic bias, rather than deliberate actions to undermine individual or group security.

To address this gap, this book aims to provide a comprehensive analysis of the interaction of AI and security, focussing on intentional use for harmful outcomes.

AI systems possess significant capabilities that go beyond human performance in various tasks. Recent advancements have demonstrated AI systems' ability to excel in image recognition, game-playing, speech recognition, language comprehension, and vehicle navigation. These advancements have significant implications for security, as they can be applied in autonomous weapon systems, disinformation campaigns, and surveillance.

AI systems also exhibit security-relevant properties such as efficiency, scalability, and the ability to exceed human capabilities. These properties enable AI systems to expand existing threats, introduce new threats, and alter the typical character of threats:

AI expands threat actors: Progress in AI expands the set of threat actors capable of carrying out attacks, increases the frequency and scale of attacks, and extends the set of potential targets. AI systems' efficiency and scalability allow for more effective attacks, such as spear phishing, which can be tailored to specific individuals or groups. The increasing anonymity provided by AI systems may also lead to a greater willingness to carry out attacks.

Additionally, the diffusion of AI systems and the declining cost of hardware facilitate the expansion of existing threats. For example, the proliferation of affordable consumer drones enables non-state actors to launch aerial attacks.

AI introduces new threats: AI advancements enable attackers to carry out attacks that were previously infeasible for humans. The ability to mimic voices and generate synthetic content allows for the spread of disinformation and impersonation. AI systems can also control robots and malware in ways that surpass human capabilities, leading to new attack vectors.

Moreover, the vulnerabilities of AI systems provide opportunities for attacks that exploit these weaknesses. Attacks targeting self-driving cars or centralised servers controlling autonomous weapon systems demonstrate the potential exploitation of AI vulnerabilities.

AI alters the typical character of threats: The combination of AI capabilities and security-relevant properties leads to a shift in the typical character of threats. Attacks supported by AI are expected to be highly effective, finely targeted, difficult to attribute, and exploit vulnerabilities in AI systems. The efficiency and scalability of AI systems enable more effective attacks, while increasing anonymity facilitates difficult-to-attribute attacks.

Exploitation of AI vulnerabilities introduces a new dimension to the threat landscape.

The Security Landscape has changed. The expansion of existing threats, introduction of new threats, and alteration of the typical character of threats necessitate proactive responses. To address the potential risks associated with AI, it is crucial to develop interventions that assess risks, protect potential victims, and prevent malicious actors from accessing and deploying dangerous AI capabilities.

As AI continues to advance, its impact on the cybersecurity landscape will continue to evolve. This book provides insights into the potential risks and implications of the malicious use of AI. However, further research is needed to explore additional interventions and strategies that can effectively address these challenges.

It is essential to strike a balance between the benefits and risks of AI, ensuring that the development and deployment of AI technologies prioritise security and accountability. By understanding the potential threats and taking proactive measures, we can harness the power of AI while minimising the risks it presents to individuals, organisations, and society as a whole. AI is having huge implications for cybersecurity:

Automation of social engineering attacks: One of the primary challenges in combating cyber threats is the rise of social engineering attacks. Hackers exploit human vulnerabilities, tricking individuals into divulging sensitive information or clicking on malicious links. With the advent of AI, these attacks can be automated and personalised on an unprecedented scale. AI-powered systems can analyse vast amounts of data to generate custom malicious websites, emails, or links that are tailored to the victim's preferences and contacts. By impersonating trusted sources and mimicking their writing styles, AI chatbots can engage in prolonged conversations, establishing trust and increasing the chances of successful exploitation.

Automation of vulnerability discovery: AI can significantly speed up the process of discovering vulnerabilities in software and systems. By analysing historical patterns of code vulnerabilities, AI algorithms can identify and exploit new weaknesses quickly. This automation allows security professionals to proactively address vulnerabilities and patch them before they can be exploited by malicious actors.

More sophisticated automation of hacking: Traditionally, hackers have relied on human intuition and manual techniques to select targets, evade detection, and respond to changes in the target's behaviour. However, AI brings a new level of sophistication to hacking. AI-powered tools can autonomously select and prioritise targets, adapt to evolving defences, and outperform human hackers in terms of speed and efficiency. While fully autonomous AI hacking tools capable of surpassing human capabilities are still a future possibility, the potential is undeniable.

Human-like denial-of-service attacks: Denial-of-service (DoS) attacks can cripple online services by overwhelming them with traffic, rendering them inaccessible to legitimate users. AI can be leveraged to orchestrate massive crowds of autonomous agents that mimic human behaviour patterns, such as click patterns and website navigation. By simulating genuine user activity, these AI-powered agents can saturate online services, effectively denying access to legitimate users and potentially compromising the system's security.

The use of AI in cybersecurity offence has the potential to increase the number, scale, and diversity of attacks that can be conducted. Conversely, AI-enabled defences are also being developed to mitigate these threats. However, further technical and policy innovations are necessary to ensure that the impact of AI on digital systems is net beneficial.

Cybersecurity has become a critical concern as IT systems have evolved into complex and sprawling networks. Many of these systems are under-maintained and insecure, creating opportunities for automation using AI.

The rise of cybercrime and targeted attacks from state actors has highlighted the need for advanced cybersecurity measures. Malicious actors employ various techniques such as DDoS attacks, malware, phishing, and ransomware. The use of AI in cyber defence, particularly in areas like anomaly and malware detection, has already shown promise in mitigating these threats.

While AI is currently employed primarily in defensive cybersecurity measures, there is growing concern about its potential use for offensive purposes. White hat researchers have conducted experiments using AI to identify vulnerabilities and propose solutions. However, the rapid progress in AI suggests that cyberattacks leveraging ML capabilities may soon become a reality. Claims of AI-based attacks have already been made, though publicly documented evidence is lacking. Nevertheless, the co-evolution of AI and cybersecurity calls for proactive preparation to counter future attacks.

Governments worldwide are recognising the importance of AI in cybersecurity. The integration of AI systems into military strategies and operations is already underway, with the US Department of Defence implementing its vision of a "Third Offset" strategy. Foundational research is being conducted to expand the capabilities of AI systems. For instance, DARPA hosted the Cyber Grand Challenge contest, where human researchers competed to develop autonomous AI systems capable of attacking and defending against other systems. As AI cybersecurity capabilities improve, governments must address the potential risks and implications of AI in their policies and regulations.

The application of AI to cybersecurity has the potential to significantly alter the threat landscape. AI can enable larger-scale and more diverse attacks by leveraging its learning capabilities. For example, AI can be used to create tailored phishing campaigns based on user interests or automate software vulnerability discovery. Attackers are likely to exploit the adaptability of AI systems to craft sophisticated attacks that current defences are ill-prepared for. The use of large datasets and the ability to vary attack details for each target can give attackers an advantage over traditional defence mechanisms. The coexistence of AI and cybersecurity necessitates the development of robust AI-based defences to keep pace with evolving threats.

Addressing cyber risks requires interventions at multiple points of control to enhance security. Consumer awareness plays a crucial role in spotting and preventing attacks. Governments and researchers contribute to cybersecurity through legislation, norms, and responsible vulnerability disclosure. Industry centralisation, particularly in spam filtering and network monitoring, aids defence efforts. Technical cybersecurity defences, such as endpoint security and anomaly detection, are also vital. However, there is a need for further development and improvement in AI-based defence technologies. The collaboration of hardware manufacturers, distributors, and robot users is essential in implementing regulations and enforcing security measures. Physical defences, such as detection systems and interception methods, can provide protection against physical attacks involving AI.

Physical Security is equally important to digital security. AI enables the customisation and equipping of robots with dangerous payloads, making them potential tools for physical attacks. The availability of open-source frameworks and software components further facilitates malicious intentions. The adaptability and long-duration operation of AI-controlled robots increase the potential for large-scale attacks and pose challenges for defence mechanisms. The intersection of cybersecurity and physical security introduces additional complexities, as the Internet of Things (IoT) and cyber-physical systems create vulnerabilities that can be exploited by AI. The emergence of AI-specific vulnerabilities, such as adversarial examples, adds to the complexity of securing autonomous systems. AI can play a significant role in protecting physical assets and preventing attacks on critical infrastructure:

Terrorist repurposing of commercial AI systems: Commercial AI systems, such as drones and autonomous vehicles, can be repurposed by terrorists to carry out harmful acts. By weaponising these AI systems, terrorists can deploy explosives or cause deliberate crashes, posing a significant threat to public safety. The ability of AI systems to operate autonomously and adapt to changing conditions makes them attractive tools for malicious actors.

Endowing low-skill individuals with high-skill attack capabilities: AI-enabled automation has the potential to reduce the expertise required to execute sophisticated attacks. For example, self-aiming, long-range sniper rifles can be equipped with AI capabilities, enabling individuals with limited training to carry out precision attacks. This democratisation of high-skill attack capabilities poses a significant challenge for security forces and underscores the need for advanced countermeasures.

Swarming attacks: Coordinated attacks involving distributed networks of autonomous robotic systems, known as swarming attacks, pose a significant threat to physical security. These autonomous systems can collaborate at machine speed, providing ubiquitous surveillance and executing rapid, synchronised attacks. Swarming attacks can overwhelm defences and compromise critical infrastructure, necessitating the development of robust countermeasures.

Attacks further removed in time and space: Autonomous operation allows physical attacks to be carried out remotely, even in environments where direct communication with the system is not possible. This increased distance between the attacker and the target reduces the risk of detection and attribution, making it more challenging to prevent and investigate such attacks.

Mitigating the risks associated with physical attacks involving AI requires interventions at various points of control. Hardware manufacturers and distributors can implement standards and regulations to ensure the security of robots and drones. Governments play a crucial role in establishing laws and norms related to lethal autonomous weapon systems. Physical defences, such as detection systems and interception methods, are essential in safeguarding against physical harm. Additionally, payload control, registration requirements, and the development of norms for responsible use contribute to physical security.

Political Security will be impacted by AI as it is a double-edged sword. While it can be harnessed for positive purposes, such as enhancing intelligence capabilities, it can also be exploited to suppress dissent and manipulate public opinion:

State use of automated surveillance platforms to suppress dissent: Governments can leverage AI to automate image and audio processing, enabling mass surveillance and intelligence collection on an unprecedented scale. This can be employed to suppress dissent, monitor citizens, and manipulate public discourse. The potential for abuse raises concerns about privacy and civil liberties.

Fake news reports with realistic fabricated video and audio: AI-powered technologies can generate highly realistic videos and audio, making it increasingly challenging to distinguish between genuine and fabricated content. Malicious actors can exploit this technology to create fake news reports featuring state leaders making inflammatory comments they never actually made. These fabricated reports can have far-reaching consequences, sowing division and undermining trust in institutions.

Automating influence campaigns: AI-enabled analysis of social networks allows malicious actors to identify key influencers and target them with offers or disinformation. By leveraging AI algorithms, influence campaigns can be automated, amplifying their impact and reaching a broader audience. This manipulation of public opinion poses a significant challenge for democratic societies.

Denial-of-information attacks: Bot-driven information-generation attacks can flood information channels with noise, making it difficult for users to acquire accurate and reliable information. By overwhelming the information space with false or distracting information, malicious actors can manipulate user behaviour and shape public discourse.

Manipulation of information availability: Media platforms' content curation algorithms can be utilised to drive users towards or away from certain content, influencing their beliefs and behaviours. This manipulation of information availability can have far-reaching implications, shaping public opinion and potentially undermining democratic processes.

The rapid growth of AI technology has led to a wide range of potential applications, both beneficial and malicious. AI systems can be deployed to automate attacks, exploit vulnerabilities, and manipulate data, posing significant risks to individuals, organisations, and even nations. The dual-use nature of AI, where the same technology can be used for both legitimate and malicious purposes, presents unique challenges for cybersecurity.

Collaboration is vital in addressing the challenges posed by AI. This requires close collaboration between technical researchers, policymakers, and other stakeholders. Policymakers must work hand-in-hand with technical experts to understand the potential risks and develop appropriate regulations that do not hinder research progress. Likewise, researchers and engineers in AI should take responsibility for the dual-use nature of their work, considering the potential for misuse and actively reaching out to relevant actors when harmful applications are foreseeable.

Promoting responsible use is important. A culture of responsibility should be fostered among AI researchers and organisations employing AI technology. Education and training programmes should be developed to raise awareness about the ethical and responsible use of AI. Ethical statements and standards can guide AI research, ensuring that considerations of security and potential harm are integrated into the development process. Collaboration with stakeholders from different sectors, including civil society, national security experts, and ethicists, can help ensure that diverse perspectives are considered in shaping AI policies.

To effectively address the risks associated with AI and cybersecurity, there are priority areas that require further research and analysis:

Learning from the cybersecurity community: As AI-based systems become more widespread, the potential impacts of cybersecurity incidents grow exponentially. AI is crucial to cybersecurity for several reasons. First, increased automation allows for greater control over physical systems, making them more vulnerable to attacks. Second, successful attacks on AI-based systems can expose the algorithms and trained models used, enabling further exploitation. Third, the use of AI in cyberattacks allows for highly sophisticated and scalable attacks.

To mitigate these risks, it is essential to learn from and collaborate with the cybersecurity community. Best practices from cybersecurity, such as red teaming exercises to identify and fix vulnerabilities, should be adopted in the AI context. Formal verification methods can be explored to ensure the security and robustness of AI systems. Responsible disclosure procedures for AI vulnerabilities should also be established to facilitate confidential reporting and timely fixes.

Exploring openness models: The prevailing norm in AI research is openness, with researchers sharing their findings to promote collaboration and progress. However, the potential misuse of AI technology raises questions about the extent of openness. It may be necessary to abstain from or delay publishing certain findings related to AI for security reasons. Pre-publication risk assessments can help determine the appropriate level of openness in specific areas of concern.

Different openness models should be explored, including central access licensing models that allow widespread use of AI capabilities while reducing the risk of malicious use. Sharing regimes can be established to selectively share research results among trusted organisations, ensuring responsible dissemination of knowledge. Drawing lessons from other dual-use technologies, such as biotechnology, can provide insights into the development of effective openness models for AI.

Fostering a culture of responsibility: AI researchers and organisations employing AI technology have a unique role in shaping the security landscape. It is crucial to promote a culture of responsibility and ethics within the AI community. Education and training programmes should incorporate ethical and socially responsible use of AI technology. Ethical statements and standards can guide AI research, ensuring that the potential for harm is considered at every stage.

Whistleblowing measures can be established to encourage the reporting of vulnerabilities and potential exploits. Engaging with a diverse range of stakeholders, including civil society, national security experts, and the general public, can help foster a comprehensive dialogue on the appropriate uses of AI technology.

AI has experienced significant technological advancements in recent years, leading to breakthroughs in machine perception, optimisation, and generative algorithms. These advancements have sparked discussions about the possibility of achieving artificial general intelligence (AGI), which refers to AI systems that possess human-like cognitive capabilities. While AGI remains a topic of both excitement and anxiety, AI is already being recognised as a powerful tool for addressing critical challenges in various domains.

With the rapid pace of technological advancement, it is essential to consider the vulnerabilities, threats, and risks associated with AI. To effectively manage these risks, it is crucial to understand how AI manifests in different applications. We can categorise AI into three broad categories: AI as a component or subsystem, AI as human augmentation, and AI with agency. Each category presents unique threats and risks that require specific regulatory considerations.

Before diving into regulatory considerations, it is essential to understand the threats and risks associated with AI. AI can be categorised into three broad applications: AI as a subsystem, AI as human augmentation, and AI with agency.

AI can be viewed as a subsystem. AI is often embedded within software systems, performing functions such as machine perception and optimisation. However, these AI subsystems are vulnerable to various threats, including data poisoning attacks and adversarial input attacks. To mitigate these risks, it is necessary to identify and protect against new and existing threats through AI-specific assurance measures and cybersecurity practices.

Regulations should require AI components to meet software assurance requirements and AI-specific assurance requirements. These requirements can be developed based on validated AI assurance frameworks. Additionally, regulated industries should develop response plans based on the NIST AI Risk Management Framework (RMF) or alternative AI assurance approaches that align with their specific sector.

Human Augmentation is another consideration. AI can also be used to augment human capabilities, enabling individuals to operate on a larger scale. While this has the potential to increase productivity and improve performance in various fields, it also poses risks. Bad actors can leverage AI to enhance their adversarial capabilities, especially in cyber operations and the spread of mis/disinformation. Regulatory measures should focus on system auditability and holding individuals accountable for misusing AI to cause harm. This auditability enables the documentation of intent and the execution of malicious actions with AI. Legal frameworks should differentiate between intentional misuse and accidents, ensuring appropriate levels of transparency to detect and mitigate intentional AI misuse.

Agency is another consideration with AI. There is growing concern about scenarios where sophisticated AI operates as an independent, goal-seeking agent. While this may seem like science fiction, it is essential to consider the potential risks associated with AI with agency. One example is the development of autonomous malware or AI models with malicious intent. Regulatory frameworks should address the risks posed by AI with agency, particularly in critical infrastructure systems, and consider automated red teaming as a means to counter these risks. Regulatory frameworks should address the risks introduced by AI with agency, particularly in safety-critical cyber-physical systems. Critical infrastructure plans should consider the increased risk due to AI-enabled scale and speed and explore the use of automated red teaming for countering these risks. Additionally, federal funding should support research and development efforts to create common vocabulary and frameworks for AI alignment and differentiate between appropriate research and bad actors.

A regulatory framework for AI security must contain several key components. Based on the vulnerabilities, threats, and risks associated with AI, we propose several elements of a regulatory framework to ensure AI security. These elements address AI as a component or subsystem, AI augmenting human capabilities, and AI with agency.

Regulatory considerations for AI as a component or subsystem:

1. AI components should satisfy both software assurance requirements and AI-specific assurance requirements. This ensures that AI subsystems undergo rigorous testing, validation, and verification processes to identify and mitigate vulnerabilities.

2. Regulated industries should develop response plans based on the National Institute of Standards and Technology's (NIST) AI Risk Management Framework (RMF). If the NIST framework is deemed insufficient, alternative AI assurance approaches should be identified.

3. The regulation should account for and mitigate risks stemming from component interactions within AI systems. Dysfunctional interactions among AI components can lead to system failures, making it necessary to establish testing standards and assurance management requirements for component interactions.

4. AI regulation should be context-specific and leverage existing regulatory frameworks in regulated sectors or industries where AI is deployed. Existing regulations can address performance standards, oversight, and legal responsibilities while accommodating the unique concerns of AI-enabled components.

5. Industry regulators should conduct continuous regulatory analysis of individual use cases to tailor appropriate regulations to match the properties of each use case. This approach avoids a one-size-fits-all approach and enables regulatory flexibility in a rapidly evolving technology landscape.

6. Industry regulators should promote trusted information-sharing mechanisms to support regulatory analysis. Establishing a common vocabulary and frameworks for AI alignment can facilitate pattern detection and identification of plausible actions for mitigating risks.

Regulatory considerations for AI implementations that aim to augment human capabilities:

7. AI regulations should require system auditability to hold individuals accountable for misusing AI to cause harm. System auditability enables the documentation of intent and execution of ill intent with AI, strengthening legal frameworks for deterrence.

8. Legal frameworks should scale accountability with the risk associated with intentional and harmful AI misuse. Increased legal penalties should be tied to higher degrees of harm caused by intentional AI misuse, while differentiation between intentional acts and "AI accidents" should be clarified.

9. AI regulation should provide appropriate levels of transparency into AI applications to an objective third party and/or the public. Transparent monitoring of AI applications can help detect and mitigate intentional AI misuse, particularly in activities that have far-reaching consequences, such as elections and the spread of mis/disinformation.

Regulatory considerations for AI implementations that have agency:

10. Federal government critical infrastructure plans should address the increased risk posed by AI-enabled scale and speed. Assessing and strengthening safety-critical cyber-physical systems vulnerable to AI threats is crucial. Automated red teaming, including AI-enabled capabilities like CALDERA, can help counter these risks.

11. Increased federal funding should be allocated to create a common vocabulary and frameworks for AI alignment. These resources can guide future research and ensure that AI advancements align with societal needs and values.

12. Regulation and legal frameworks should differentiate between appropriate research with risk mitigations and bad actors. Both unintentional and intentional harm caused by AI systems should be addressed, imposing penalties on researchers or individuals who give AI systems goals that lead to harm.

By implementing these regulatory elements, we can establish a more comprehensive framework that addresses the risks and vulnerabilities associated with AI security. It is crucial to strike a balance between promoting innovation and protecting against potential harms, ensuring that AI technology develops in a safe and responsible manner.

A sensible regulatory framework for AI security is essential to address the threats and risks associated with AI. By considering the unique characteristics of AI as a subsystem, AI augmenting human capabilities, and AI with agency, we can establish regulations that promote the safe and secure use of AI. These regulations should prioritise system auditability, accountability, transparency, and the differentiation between intentional and unintentional harm. With the right regulatory measures in place, we can harness the potential of AI while safeguarding against its risks.

NOTE

1 https://www.ibm.com/downloads/cas/V2EZNEN5.

https://orcid.org/0009-0005-0854-6213

CHAPTER 14

Ethical Concerns with Artificial Intelligence

ARTIFICIAL INTELLIGENCE (AI) has rapidly emerged as a transformative technology, revolutionising various industries and aspects of our daily lives. With its ability to simulate human intelligence and perform tasks that traditionally required human cognition, AI has demonstrated immense potential. However, this remarkable advancement also comes with a set of ethical concerns that need to be carefully addressed.

As AI continues to evolve and integrate into our society, ethical concerns have become more prominent. One significant concern revolves around the issue of privacy and data security.

With the vast amount of data being collected and analysed to train AI algorithms, there is a growing risk of unauthorised access and misuse of personal information. Striking the right balance between utilising data for AI advancements and safeguarding individual privacy becomes crucial.

The rapid growth of AI technology has led to an exponential increase in data collection. This data often includes sensitive personal information that can be exploited if not protected adequately. As AI systems become more sophisticated and capable of processing massive amounts of data, it becomes crucial to establish robust security measures and regulations to ensure the privacy of individuals. Implementing strict data protection protocols, encryption techniques, and anonymisation methods can help mitigate the risks associated with privacy breaches and unauthorised access.

DOI: 10.1201/9781003502708-15

Another ethical concern in the field of AI is the potential for bias and discrimination embedded within the algorithms. AI systems are trained on vast datasets, which can inadvertently include biases present in the real world. If these biases are not identified and addressed during the development phase, AI algorithms can perpetuate and amplify existing societal biases. This can result in unfair and discriminatory outcomes, affecting marginalised communities. It is essential to develop AI algorithms that are transparent, accountable, and regularly audited to ensure fairness and mitigate any potential biases.

As AI technology continues to advance, there are growing concerns about its impact on the job market and employment. AI has the potential to automate various tasks and roles traditionally performed by humans, leading to job displacement. While AI can enhance productivity and efficiency, it is crucial to find a balance that minimises negative consequences for the workforce. This can be achieved by upskilling and reskilling programmes, providing opportunities for workers to transition into new roles that leverage their unique human capabilities.

The ability of AI systems to make autonomous decisions raises significant ethical concerns. When AI is entrusted with decision-making authority in areas such as healthcare, criminal justice, and finance, there is a need for transparency, accountability, and human oversight. Ensuring that AI systems are designed to align with ethical principles and adhere to established guidelines becomes imperative. Human involvement in critical decision-making processes, along with clear explanations of AI rationale, can help build trust in these systems.

Striking a balance between addressing ethical concerns and fostering the advancement of AI is essential. Ethical considerations must be integrated into the development lifecycle of AI systems, starting from the design phase. This includes robust ethical frameworks, regular audits, and continuous monitoring of AI systems to ensure compliance with ethical standards. Collaboration between industry, academia, and policymakers is vital to establishing guidelines and best practices for AI development that prioritise ethical considerations.

Recognising the ethical challenges posed by AI, regulatory efforts are underway to establish guidelines and frameworks for ethical AI development. Governments and organisations worldwide are working towards creating a regulatory environment that promotes responsible and ethical

AI practices. These efforts include the establishment of ethical review boards, certification programmes, and guidelines for data usage and algorithm transparency.

Encouraging collaboration between stakeholders and fostering public discourse ensures that regulatory efforts are comprehensive and consider diverse perspectives.

The rapid advancement of AI brings with it a set of ethical concerns that must be carefully addressed. Privacy and data security, bias and discrimination, impact on employment, and the potential for autonomous decision-making are among the key concerns. Balancing these concerns with the progress and potential of AI requires a multi-faceted approach involving collaboration, transparency, and regulatory efforts. As we navigate the future of AI, it is crucial to prioritise ethical considerations and ensure that technology benefits society as a whole.

AI has emerged as a transformative technology with the potential to revolutionise various industries and improve the quality of life for people around the world. However, along with its immense benefits, AI also presents a range of ethical dilemmas that need to be addressed. Mankind will need to explore the ethical considerations associated with the development and deployment of AI, and discuss how we can strike a balance between innovation and responsibility.

AI holds tremendous potential to enhance productivity, efficiency, and decision-making across multiple sectors. From healthcare and transportation to finance and education, AI has the power to revolutionise these industries by automating tasks, analysing complex data, and providing valuable insights. However, there are also concerns surrounding AI that need to be carefully examined.

One of the primary concerns with AI is its potential to replace human jobs. As AI systems become more sophisticated, there is a fear that they will eliminate the need for human workers, leading to widespread unemployment. Another concern is the bias and discrimination that can be embedded in AI algorithms. If not properly addressed, AI systems can perpetuate existing social inequalities and reinforce discriminatory practices.

As AI continues to advance, it is crucial to consider the ethical implications of its development and deployment. Privacy is a significant concern when it comes to AI, as it often involves the collection and analysis of vast amounts of personal data. Safeguarding this data and ensuring that it is used responsibly is essential to protect individuals' privacy rights.

Another ethical consideration is the accountability of AI systems. When AI makes decisions that impact human lives, it is crucial to have mechanisms in place to hold these systems accountable for their actions. Transparency is also vital, as AI algorithms can be complex and difficult to understand. Ensuring transparency in AI systems will enable users to have a clear understanding of how decisions are being made.

Finding the right balance between innovation and responsibility is crucial in the development and deployment of AI technology. While innovation drives progress and can bring about significant benefits, it must be guided by ethical considerations to avoid potential harm.

Responsible innovation entails taking into account the potential risks and consequences of AI systems and integrating ethical principles throughout the development process.

To achieve this balance, it is essential to involve a diverse range of stakeholders, including ethicists, policymakers, technologists, and end-users, in the decision-making process. By incorporating different perspectives, we can ensure that AI technology is developed in a way that aligns with societal values and addresses the concerns of various stakeholders.

Addressing the ethical dilemmas of AI requires collaboration between governments, industry leaders, and other relevant stakeholders. Governments play a vital role in setting regulations and standards that govern the development and deployment of AI technology. These regulations should aim to protect individual rights, ensure fairness, and prevent the misuse of AI systems.

Industry leaders also have a responsibility to prioritise ethical considerations in AI development. By adopting ethical guidelines and integrating ethical principles into their AI systems, companies can promote responsible innovation and build trust with their users and the public. Collaboration between governments and industry is crucial to establish a comprehensive framework that addresses the ethical challenges of AI.

To ensure responsible AI development, it is essential to follow best practices that prioritise ethics and accountability. Firstly, developers should strive for transparency by documenting the design and decision-making process of AI systems. This transparency allows for scrutiny and helps identify any biases or errors in the algorithms.

Secondly, incorporating diverse datasets and involving a wide range of voices in the development process can help mitigate bias and ensure fairness. By proactively addressing biases and ensuring inclusivity, AI systems can be developed in a way that benefits all individuals and avoids discrimination.

Lastly, continuous monitoring and evaluation of AI systems are crucial. Regular audits and assessments can help identify any unintended consequences or biases that may have emerged over time. By regularly reviewing and updating AI systems, developers can ensure that they align with ethical standards and address societal concerns.

Transparency and accountability are fundamental principles that should underpin AI systems to address ethical dilemmas. Transparency involves making the decision-making process of AI systems understandable and accessible to users and stakeholders. This transparency enables individuals to have confidence in the decisions made by AI and ensures that potential biases or errors can be identified and corrected.

Accountability, on the other hand, involves holding AI systems and their developers responsible for the outcomes of their decisions. Establishing clear lines of responsibility and mechanisms for redress is crucial to ensure that individuals affected by AI systems have recourse in the event of harm or unfair treatment.

To guide ethical decision-making in AI systems, the development of ethical frameworks is imperative. These frameworks provide a set of principles and guidelines that developers can follow to ensure ethical behaviour in AI technology. Ethical frameworks should incorporate principles such as fairness, transparency, privacy, and accountability.

One example of an ethical framework is the "Ethics of AI and Robotics" developed by the European Commission. This framework emphasises the importance of human oversight, transparency, and accountability in AI systems. Similarly, the Institute of Electrical and Electronics Engineers (IEEE) has developed a set of ethical guidelines for AI developers that promote fairness, transparency, and the protection of individual rights.

As AI continues to advance, it is crucial to address the ethical dilemmas associated with its development and deployment. Striking a balance between innovation and responsibility requires collaboration between various stakeholders, including governments, industry leaders, and ethicists. By following best practices, promoting transparency and accountability, and developing ethical frameworks, we can ensure that AI technology is developed in a way that benefits society while upholding ethical standards. It is only through these efforts that we can harness the full potential of AI while minimising its potential harms.

As AI continues to advance and play a prominent role in various industries, concerns surrounding its ethical implications and potential negative impacts have become more prevalent. We can consider the real-life case studies that highlight some of the most significant concerns related to AI. By examining these cases, we can shed light on the ethical, bias and discrimination, and job displacement concerns associated with the use of AI technology.

One of the major concerns surrounding AI is its ethical implications. As AI systems become more autonomous and capable of making decisions without human intervention, questions arise about the accountability and transparency of these systems. For instance, in the case of autonomous vehicles, ethical dilemmas arise when the AI system has to make split-second decisions that may involve sacrificing the safety of passengers or pedestrians. This raises questions about who should be held responsible for any harm caused by AI systems.

Another ethical concern is the potential for AI systems to be used for malicious purposes. For example, AI-powered surveillance systems can be used to invade people's privacy or discriminate against certain individuals or groups. It is crucial to establish ethical guidelines and regulations to ensure that AI technology is used responsibly and in a manner that respects fundamental human rights.

Bias and discrimination are significant concerns when it comes to AI technology. AI systems are trained using vast amounts of data, and if this data is biased or reflects societal prejudices, the AI system can perpetuate and amplify these biases. This can lead to unfair outcomes in various areas, such as hiring processes or criminal justice systems.

A notable case study highlighting this concern is algorithmic bias in hiring processes. AI-powered algorithms used in recruitment can inadvertently discriminate against certain demographics, such as women or people of colour. This occurs when the algorithms are trained on historical data that reflects biased hiring practices. It is essential to address these biases and ensure that AI systems are designed to be fair and unbiased.

The rapid advancement of AI technology also raises concerns about job displacement. As AI systems become more capable of performing complex tasks, there is a fear that many jobs will be automated, leading to unemployment and economic inequality. While AI has the potential to improve efficiency and productivity, it is crucial to consider the impact on the workforce and implement measures to reskill and upskill workers.

Facial recognition technology and privacy concerns: Facial recognition technology has gained significant attention in recent years, but it also raises serious privacy concerns. This technology can be used to identify individuals in real-time, potentially enabling mass surveillance without consent. For example, in China, facial recognition technology is used extensively for law enforcement purposes, raising concerns about privacy and the abuse of power. It is crucial to establish clear regulations and safeguards to ensure that facial recognition technology is used responsibly and with respect for privacy rights.

Algorithmic bias in hiring processes: Algorithmic bias in hiring processes is a real concern. AI-powered algorithms used in recruitment can perpetuate biases present in historical hiring data, leading to discrimination against certain demographics. For instance, if the algorithm is trained on data that shows a bias against women in certain job roles, it may continue to discriminate against qualified female candidates. To address this concern, organisations must critically evaluate the algorithms they use and implement measures to mitigate bias.

Autonomous vehicles and ethical concerns: Autonomous vehicles present a unique set of ethical concerns. These vehicles rely on AI systems to make split-second decisions that can have life-or-death consequences. For example, in a situation where an autonomous vehicle has to choose between hitting a pedestrian or swerving to avoid them but potentially harming its passengers, ethical dilemmas arise. It is crucial to establish clear ethical guidelines for AI systems in autonomous vehicles to ensure the safety of all individuals involved.

AI-powered chatbots and customer privacy: AI-powered chatbots have become increasingly popular in customer service, but their use raises concerns about customer privacy. Chatbots have access to a vast amount of personal data, and if not properly secured, this information can be misused or compromised. Organisations must prioritise data privacy and security when implementing AI-powered chatbots and ensure that appropriate measures are in place to protect customer information.

AI-powered surveillance systems: AI-powered surveillance systems have the potential to infringe upon individuals' privacy rights and enable mass surveillance. These systems can analyse vast amounts of data, including facial recognition and behavioural patterns, to track and monitor individuals. While surveillance systems have their benefits in terms of public safety, it is crucial to have clear regulations and oversight to prevent abuse and protect individual privacy.

To address the concerns surrounding AI, a multi-faceted approach is necessary. First and foremost, ethical guidelines and regulations must be established to ensure the responsible use of AI technology. This includes transparency in AI decision-making processes and mechanisms for accountability.

Additionally, organisations must actively work to mitigate bias and discrimination in AI systems. This involves critically evaluating the data used to train AI algorithms and implementing measures to ensure fairness and diversity. It is also crucial to invest in reskilling and upskilling programmes to mitigate job displacement and ensure that the workforce is prepared for the future.

AI has the potential to revolutionise various industries, but it also raises significant concerns that must be addressed. By examining real-life case studies, we have highlighted the ethical, bias and discrimination, and job displacement concerns associated with AI technology. It is crucial for policymakers, organisations, and individuals to work together to ensure that AI is developed and used in a responsible and ethical manner. Only through proactive measures can we harness the benefits of AI while mitigating its potential negative impacts.

https://orcid.org/0009-0005-0854-6213

CHAPTER 15

Disinformation and Manipulation

ARTIFICIAL INTELLIGENCE (AI) has modernised various industries, from healthcare to finance. However, its impact on disinformation is an area that requires careful consideration.

Disinformation, the deliberate spread of false or misleading information, has become a significant concern in our increasingly digitalised world. AI technology has played a pivotal role in the rapid dissemination of disinformation, posing serious challenges for society. We must explore the concerns surrounding AI and disinformation, examine the role of AI in spreading disinformation, and delve into examples of AI-enabled disinformation campaigns.

As AI technology continues to advance, concerns about its potential misuse and manipulation have arisen. One major concern is the ability of AI to generate highly convincing fake content, including realistic images, videos, and audio recordings. These advancements make it increasingly difficult to distinguish between real and fake information. Moreover, the speed and scale at which AI can spread disinformation is unprecedented, making it challenging for human fact-checkers to keep up. This raises questions about the reliability of information sources and undermines trust in traditional media.

Another concern is the potential for AI algorithms to reinforce existing biases and echo chambers. AI systems learn from vast amounts of data, which can be biased or skewed. If these biases are not addressed, AI can

DOI: 10.1201/9781003502708-16

inadvertently perpetuate and amplify disinformation by selectively presenting information that aligns with users' existing beliefs. This creates an echo chamber effect, where individuals are exposed to a limited range of perspectives and are less likely to critically evaluate information.

AI technology has facilitated the creation and dissemination of disinformation on an unprecedented scale. AI-powered bots and algorithms can automate the process of generating and spreading false information, allowing disinformation campaigns to reach a wider audience in a shorter amount of time. These AI-generated narratives can manipulate public opinion, influence elections, and sow discord within societies.

AI's role in spreading disinformation extends beyond social media platforms. Deepfake technology, a form of AI-generated synthetic media, has the potential to deceive individuals by creating highly realistic videos or audio recordings of people saying or doing things they never actually did. Deepfakes can be used to spread false information, defame individuals, or even incite violence. The combination of AI and disinformation poses a significant threat to our information ecosystem and democratic processes.

Several high-profile cases illustrate the impact of AI-enabled disinformation campaigns. One notable example is the use of AI-powered bots during the 2016 US presidential election.

These bots were designed to amplify divisive messages, spread misinformation, and manipulate public opinion. The sheer volume and speed at which these bots disseminated false information had a significant impact on the election discourse, polarising voters and undermining trust in the democratic process.

Another example is the proliferation of deepfakes, particularly in the realm of politics. In 2019, a deepfake video of Nancy Pelosi, the Speaker of the United States House of Representatives, went viral on social media. The video was manipulated to make Pelosi appear drunk, raising concerns about the potential for deepfakes to manipulate public perception and undermine the credibility of public figures.

Addressing AI-generated disinformation is a complex and multifaceted challenge. One major obstacle is the rapid evolution of AI technology itself. As AI algorithms become more sophisticated, disinformation campaigns can adapt and evolve accordingly. This cat-and-mouse game between AI-generated disinformation and countermeasures requires constant vigilance and technological advancements.

Another challenge is the sheer volume of disinformation circulating online. AI technology enables the rapid creation and dissemination of vast amounts of false information, overwhelming traditional fact-checking methods. Human fact-checkers simply cannot keep up with the speed and scale of AI-generated disinformation. This necessitates the development of AI-powered tools and algorithms to combat disinformation effectively.

Additionally, the global nature of disinformation campaigns poses jurisdictional challenges. Disinformation can originate from anywhere in the world and target any country or society. Cooperation and collaboration between tech companies, governments, and users on a global scale are essential to effectively address AI-generated disinformation.

While addressing AI-generated disinformation is a daunting task, there are solutions that can help mitigate its impact. One crucial approach is the development of advanced AI detection and verification systems. AI can be utilised to detect patterns and anomalies in online content, flagging potential disinformation for further analysis. Additionally, AI can assist in verifying the authenticity of content, helping to identify deepfakes and other manipulated media.

Collaboration between tech companies, governments, and users is paramount in the fight against AI-generated disinformation. Tech companies must prioritise the development and implementation of robust content moderation policies and AI-based detection systems.

Governments should enact legislation that holds both individuals and organisations accountable for spreading disinformation. Users, on the other hand, should be educated about the risks of disinformation and empowered to critically evaluate the information they encounter online.

Effective solutions to combat AI-generated disinformation require collaboration and cooperation between tech companies, governments, and users. Tech companies have a responsibility to develop and implement transparent content moderation policies, ensuring that their platforms are not used as vehicles for the spread of disinformation. Governments should work closely with tech companies to establish regulations and legislation that hold individuals and organisations accountable for spreading disinformation. Users, on the other hand, need to be educated about the risks of disinformation and equipped with the tools to critically evaluate information.

By working together, these stakeholders can create a more resilient information ecosystem that combats AI-generated disinformation effectively. Sharing knowledge, resources, and best practices will be key to staying ahead of rapidly evolving disinformation tactics. Regular dialogue and collaboration can help identify emerging threats and develop proactive strategies to address them.

The rapid advancement of AI and algorithms has significantly impacted various aspects of society, including the media landscape. While algorithms have the potential to enhance personalised experiences and improve content recommendation, they also pose challenges related to algorithmic bias and disinformation. Algorithmic bias refers to the systematic favouritism or discrimination that may occur in the automated decision-making processes of algorithms. In the context of media, algorithmic bias can lead to the amplification of misinformation, distortion of facts, and reinforcement of existing biases.

We should explore the issue of algorithmic bias in AI and its impact on media. By understanding the complexities of algorithmic media curation and the role of different actors within the system, we can identify the challenges associated with algorithmic personalisation and the potential consequences of algorithmic amplification. Additionally, we discuss the importance of promoting fairness and transparency in algorithmic systems and outline research directions to address algorithmic bias effectively.

Algorithmic bias refers to the inherent or learned biases that are reflected in the decisions made by algorithms.[1] These biases can arise from various sources, including biased training data, flawed algorithms, or the influence of human biases in the design and implementation of algorithms. There are several types of algorithmic bias, including:

1. *Selection bias*: This type of bias occurs when algorithms favour certain groups or individuals over others in the selection and presentation of content. It can result in the underrepresentation or exclusion of certain perspectives, leading to an imbalanced flow of information.

2. *Prejudice bias*: Prejudice bias involves the perpetuation of stereotypes or discriminatory treatment based on attributes such as race, gender, or socioeconomic status. Algorithms that exhibit prejudice bias can reinforce existing inequalities and discrimination.

3. *Confirmation bias*: Confirmation bias occurs when algorithms prioritise content that aligns with users' existing beliefs or preferences, leading to a filter bubble effect. This can limit users' exposure to diverse perspectives and contribute to the polarisation of society.

Algorithmic bias in media has significant implications for democracy, informed decision-making, and societal cohesion. Biased algorithms can perpetuate misinformation and disinformation, leading to the spread of false narratives and the erosion of trust in media sources. The filter bubble effect can create echo chambers, where individuals are only exposed to information that confirms their existing beliefs, reinforcing polarisation and exacerbating societal divisions.

Moreover, algorithmic bias can contribute to the marginalisation and underrepresentation of certain groups, leading to the amplification of systemic biases and discrimination. Media platforms have a responsibility to ensure that algorithms do not perpetuate harmful stereotypes or reinforce discriminatory practices.

Algorithmic media curation involves a complex interplay of various actors, including algorithms, users, and media creators. Understanding the dynamics between these actors is crucial for comprehending the challenges and potential solutions related to algorithmic bias.

The primary actors in algorithmic media curation are algorithms, users, and media creators. Algorithms are responsible for processing vast amounts of data and making recommendations based on predefined criteria. Users play a role in shaping algorithmic recommendations through their preferences, choices, and interactions with content. Media creators contribute to the availability of content that algorithms can curate and recommend.

Algorithms play a central role in algorithmic media curation by analysing user data, identifying patterns, and making content recommendations. While algorithms have the potential to provide personalised and relevant content, they can also introduce biases if not designed and implemented carefully. Algorithmic systems must consider ethical considerations, transparency, and accountability to ensure fair and unbiased content curation.

Users have an impact on algorithmic bias through their interactions with content and their preferences. However, users may also be influenced by algorithmic recommendations, leading to a feedback loop that reinforces existing biases. The challenge lies in striking a balance between personalisation and exposure to diverse perspectives, ensuring that users are not trapped in filter bubbles or echo chambers.

Media creators play a crucial role in algorithmic media curation by producing and publishing content. The availability of diverse and high-quality content is essential for algorithms to curate recommendations that reflect a range of perspectives. Media creators also have a responsibility to ensure that their content is unbiased and free from discriminatory practices.

Algorithmic personalisation has gained popularity in media platforms as a means to enhance user experiences and engagement. However, there are several challenges associated with algorithmic personalisation that need to be addressed to mitigate algorithmic bias.

Algorithmic personalisation can create an illusion of tailored content, giving users the perception that the recommendations align perfectly with their preferences. In reality, algorithmic personalisation relies on limited choices and predetermined criteria, which may not accurately reflect users' diverse interests and needs. Users should be aware of the limitations of algorithmic personalisation and the potential biases it may introduce.

Algorithmic personalisation faces feasibility and economic challenges. It is technically challenging to accurately capture individual preferences and deliver content that precisely matches them. Moreover, allocating resources to cater to diverse individual preferences in a highly fragmented media market may not be economically viable for media platforms.

Striking a balance between personalisation, feasibility, and economic considerations is crucial to ensure fair and unbiased content recommendations.

Algorithmic amplification refers to the process by which algorithms reinforce existing beliefs, biases, or preferences through content recommendations. This amplification can have significant implications for democratic discourse, trust in media, and societal cohesion.

Algorithmic amplification can contribute to the spread of deceptive content and misinformation. By recommending content that aligns with users' existing beliefs or preferences, algorithms can unintentionally amplify false narratives and conspiracy theories. This spiral of deceptive content can erode trust in media and hinder informed decision-making.

Algorithmic amplification can exacerbate societal divisions and polarisation. By reinforcing users' existing beliefs and limiting exposure to diverse perspectives, algorithms can contribute to the formation of echo chambers and filter bubbles. This polarisation can lead to a breakdown in trust, hinder constructive dialogue, and undermine democratic processes.

Algorithmic amplification challenges the epistemic foundations of deliberative democracy, where diverse perspectives are essential for informed decision-making. By limiting exposure to diverse viewpoints, algorithms can hinder the free exchange of ideas and inhibit the formation of well-rounded opinions. Striking a balance between personalisation and exposure to diverse perspectives is crucial for maintaining a healthy democratic discourse.

To address algorithmic bias and promote fairness in algorithmic media curation, it is essential to prioritise transparency, accountability, and user control. Educating news users about the limitations and potential biases of algorithms is crucial for empowering them to make informed choices.

News users should be provided with educational resources and tools to understand how algorithms work and how they can control their exposure to biased or deceptive content. Algorithmic literacy is essential for empowering users to navigate the media landscape critically.

Media platforms and algorithm creators must prioritise transparency and accountability in algorithmic systems. Clear explanations of how algorithms curate content, the criteria used, and the potential biases should be provided to users. Regular audits and independent assessments can help ensure fairness and prevent algorithmic bias.

Users should have more control over their algorithmic experiences. This can include options to personalise their content preferences, adjust algorithmic recommendations, or opt-out of certain algorithmic features. Informed choices should be supported by clear information about the implications and potential biases associated with algorithmic personalisation.

To effectively address algorithmic bias, further research is needed to explore various aspects of algorithmic media curation, its impact, and potential interventions. Several research directions can contribute to a better understanding of algorithmic bias and inform the development of fair and transparent algorithmic systems.

Research should move beyond the traditional focus on voluntary selective exposure and confirmation bias. Understanding how algorithmic recommendations influence users' content consumption and the potential impact on their beliefs and attitudes is crucial for addressing algorithmic bias effectively.

To comprehensively understand the effects of algorithmic bias, research should adopt a holistic approach that considers the interplay between algorithms, users, and media creators. This approach allows for a deeper understanding of how different actors contribute to algorithmic bias and its consequences.

Research should explore the long-term impact of algorithmic amplification on societal trust, polarisation, and democratic discourse.

Understanding the dynamics of algorithmic amplification over time can inform interventions and strategies to mitigate its negative consequences.

Algorithmic bias in AI presents significant challenges in the media landscape. However, by understanding the complexities of algorithmic media curation and the roles of different actors, we can develop strategies to promote fairness and transparency. Educating news users, prioritising transparency and accountability, and empowering users with more control are essential steps towards countering algorithmic bias. Further research on algorithmic bias and its impact is crucial for developing effective interventions and promoting fair and unbiased algorithmic systems in the media.

Algorithms, despite being created by humans, are often seen as objective and fair. However, this is far from the truth. AI systems are susceptible to bias due to the way they are designed, trained, and deployed. The performance of an algorithm heavily depends on the biases of the people who design them, the code they use, the data they analyse, and the way they train the models. Algorithms are not neutral; they reflect and amplify the larger prejudices of the society in which they were created.[2] This phenomenon, known as algorithmic amplification, is particularly prevalent in platforms such as Google, TikTok, Instagram, and YouTube, which are designed for data gathering, automated processing, and maximising monetisation of customer data. These platforms rely on algorithms to process vast amounts of data and drive people's decisions and behaviours in ways that may favour certain groups over others.

The vulnerability of AI systems to bias arises from various factors. Firstly, algorithms are created by people who can be influenced by their own biases and prejudices. Even well-intentioned individuals can unknowingly embed their conscious or unconscious biases into the algorithms they develop. These biases are then perpetuated and amplified by the machine learning (ML) models that power AI systems. Secondly, algorithms can operate based on secondary and non-observational data, such as synthesised data or generalised assumptions, which can introduce biases into the system. Furthermore, the data analysed or used to train ML models may not accurately reflect the diverse parameters of users, leading to biased results. In summary, the vulnerabilities of AI to bias lie in the inherent biases of the people who design and deploy algorithms, as well as in the data and assumptions used in the training process.

Algorithmic bias manifests in various ways across different platforms and domains. One common form of bias is the favouring of certain groups of users over others, resulting in unfair discriminations. For example, social media platforms often amplify sensational content to maximise user engagement and revenue. Similarly, news recommender systems may suggest negative or incorrect information, leading to the spread of misinformation and polarised opinions. Biases in algorithmic systems are not limited to social media; they are also prevalent in critical domains such as healthcare, criminal justice, and employment systems.

Biased algorithms can influence critical decisions, perpetuate social stereotypes, and exacerbate existing inequalities in society.

The causes of algorithmic bias are multifaceted and intertwined with the social, cultural, and political realities of our world. Humans play a crucial role in the development of algorithms, as they write the code, choose the data, and make decisions about how the algorithms should function. Biases can easily enter the algorithmic systems through pre-existing social and cultural values, which can influence decisions related to data gathering, filtering, coding, and analysis. These biases can then be algorithmised and automated by AI systems, leading to decisions that are systematically unfair to certain groups of people. A prime example of algorithmic bias is the use of predictive algorithms in criminal justice systems, such as the Correctional Offender Management Profiling for Alternative Sanctions (COMPAS) study.

The study revealed that the algorithms used to predict the likelihood of an inmate recommitting a crime were biased against certain racial and ethnic groups, resulting in unfair treatment and perpetuation of stereotypes.

Algorithmic bias has far-reaching consequences in society. Biased algorithms can intensify social stereotyping, reinforce inequalities, and perpetuate discrimination. For instance, speech recognition systems driven by AI have shown significant racial disparities, misunderstanding words from minority users more frequently than those from white users. Similarly, biased search engine results and social media recommendations can shape people's perceptions and reinforce existing biases. Furthermore, algorithmic bias can hinder progress in critical areas such as healthcare, where biased algorithms can lead to incorrect diagnoses or unequal access to treatment. In the employment sector, biased algorithms can perpetuate gender and racial inequalities, leading to unfair hiring practices and limited opportunities for underrepresented groups. It is crucial to recognise the pervasive nature of algorithmic bias and its potential to undermine trust in AI systems and perpetuate social injustices.

Detecting and mitigating algorithmic bias is a complex task that requires a multifaceted approach. Technical solutions alone are insufficient; addressing algorithmic bias necessitates changes in both the algorithms and the underlying social and cultural biases. To remove or mitigate bias, it is essential to consider the entire life cycle of an algorithm, from data collection to deployment. This involves ensuring diverse representation in the data used for training ML models, critically examining the assumptions and biases in the algorithms, and promoting transparency and accountability in decision-making processes. Additionally, human fact-checkers play a vital role in verifying the accuracy of information and detecting fake news and misinformation. Collaboration between humans and AI systems is crucial to ensure fair and unbiased outcomes.

Fairness and transparency are foundational principles in the design and development of algorithmic systems. Fairness refers to the equitable treatment of individuals and groups, while transparency entails openness and accountability in algorithmic decision-making. Fair and transparent algorithms foster trust, enhance user experience, and promote inclusivity.

Platforms that prioritise fairness and transparency are more likely to provide high-quality recommendations and personalised services. Open visibility and clear transparency of relevant recommendations empower users to make informed decisions and understand the underlying processes of algorithmic systems. Achieving fairness and transparency in algorithms requires ongoing efforts to address biases, improve data collection practices, and ensure that critical decisions are accountable and subject to scrutiny.

Algorithmic bias is a pressing issue that demands attention and action. The impact of biased algorithms can be far-reaching, perpetuating discrimination, reinforcing stereotypes, and exacerbating social inequalities. To address algorithmic bias, it is crucial to understand the vulnerabilities of AI systems and the factors that contribute to bias. Detecting and mitigating bias requires a comprehensive approach that involves technical solutions, changes in algorithm design, and a commitment to fairness and transparency. By striving for fair and transparent algorithms, we can build AI systems that promote inclusivity, mitigate biases, and contribute to a more equitable and just society.

As we navigate the impact of AI on disinformation, ethical considerations must be at the forefront of our discussions. AI technology, when misused, can have profound societal implications. It is crucial to develop

AI algorithms and systems that are transparent, fair, and unbiased. This includes addressing biases in training data and ensuring that AI systems are accountable and explainable.

Moreover, ethical considerations extend to the responsible use of AI in detecting and combating disinformation. Balancing the need for privacy and security with the imperative to identify and mitigate the spread of disinformation is a delicate task. Safeguarding individual rights and freedoms while protecting the integrity of our information ecosystem requires careful ethical deliberation.

While AI has played a significant role in spreading disinformation, it can also be harnessed to detect and verify false information. AI-powered algorithms can analyse patterns, linguistic cues, and contextual information to identify potential disinformation. Natural language processing (NLP) techniques can be employed to analyse the sentiment, credibility, and reliability of online content.

Furthermore, AI can assist in verifying the authenticity of media, such as images and videos, by analysing digital fingerprints, metadata, and other forensic information. These AI-powered verification systems can help distinguish between genuine and manipulated media, enabling users to make more informed judgements about the information they encounter.

As AI technology continues to evolve, its impact on disinformation will be an ongoing challenge. The rapid spread of AI-generated disinformation poses significant risks to our information ecosystem and democratic processes. However, by understanding the concerns surrounding AI and disinformation, exploring examples of AI-enabled disinformation campaigns, and implementing effective solutions, we can navigate the future of AI and disinformation.

Collaboration between tech companies, governments, and users will be essential in combating AI-generated disinformation. The development of advanced AI detection and verification systems, along with robust content moderation policies, will help mitigate the impact of disinformation. Ethical considerations should guide the development and use of AI technology, ensuring transparency, fairness, and accountability.

By addressing these challenges head-on and harnessing the potential of AI for positive change, we can create a more resilient and trustworthy information ecosystem that withstands the threats posed by AI-generated disinformation.

As AI becomes an increasingly integral part of our lives, it brings with it a host of ethical concerns. One of the most pressing issues is AI manipulation, where AI systems are intentionally designed to deceive or manipulate users for various purposes. We can explore the ethical implications of AI manipulation, its impact across different industries, and how we can navigate these challenges while ensuring accountability.

AI manipulation raises significant ethical concerns due to its potential to exploit and deceive individuals. With the ability to analyse vast amounts of data and learn from it, AI systems can be programmed to intentionally manipulate users' emotions, decisions, and behaviours. This raises questions about consent, autonomy, and privacy. For instance, social media platforms have been criticised for using AI algorithms to manipulate users' newsfeeds, leading to echo chambers and the spread of disinformation. This not only undermines democratic processes but also has real-world consequences.

The impact of AI manipulation extends beyond individual users. It has broader societal implications, including economic, political, and social consequences. For instance, AI manipulation can be used to influence financial markets, manipulate public opinion during elections, or perpetuate biases and discrimination. Furthermore, AI-powered deepfake technology can create realistic but fabricated videos and images, leading to misinformation and potential harm to individuals or organisations. As AI becomes more sophisticated, the potential for manipulation grows, highlighting the need for ethical considerations and regulation.

AI manipulation is not limited to a single industry but is prevalent across various sectors. In the healthcare industry, for example, AI systems can be manipulated to prioritise profits over patient well-being, leading to biased treatment recommendations or skewed research outcomes. In the financial sector, AI algorithms can be manipulated to exploit market trends for personal gain, leading to economic instability. In the media industry, AI manipulation can be used to generate clickbait headlines or tailored content, influencing public opinion and compromising journalistic integrity. These examples illustrate the diverse ways AI manipulation can occur and its potential consequences.

To navigate the challenges of AI manipulation, it is crucial to establish ethical guidelines and best practices for AI developers. Transparency and accountability should be prioritised, ensuring that AI systems are designed with the best interests of users and society in mind.

Developers should be encouraged to adopt a user-centric approach, focusing on the ethical implications of their creations. Additionally, interdisciplinary collaboration between AI experts, ethicists, and policymakers can help identify and address potential risks and biases in AI systems. Regular audits and third-party assessments can also provide independent verification of AI system integrity.

Accountability is key to mitigating the risks of AI manipulation. Developers should be held responsible for the ethical implications of their AI systems, ensuring that they are designed and deployed in a manner that respects user autonomy, privacy, and well-being. This may involve establishing regulatory frameworks that outline the responsibilities and liabilities of AI developers. Additionally, user education and awareness programmes can empower individuals to understand and recognise AI manipulation, enabling them to make informed decisions and protect themselves.

To ensure ethical AI development, clear guidelines need to be established. These guidelines should emphasise transparency, fairness, and accountability. Developers should be encouraged to disclose the use of AI systems and provide clear explanations of how they work. They should also strive to eliminate biases and ensure that AI systems are fair and inclusive. Regular auditing and testing should be conducted to identify and rectify any potential manipulation or bias in AI algorithms. By following these ethical guidelines, developers can contribute to a more responsible and trustworthy AI ecosystem.

AI manipulation also has legal implications, as it can lead to harm or infringe upon individuals' rights. Existing laws may need to be updated or new legislation introduced to address AI manipulation specifically. Legal frameworks should define the boundaries of AI usage and establish penalties for unethical AI manipulation. Simultaneously, laws should strike a balance between encouraging innovation and holding AI developers accountable for the consequences of their actions. Collaboration between legal experts, policymakers, and technologists is essential to create effective and enforceable regulations.

Government and regulatory bodies play a crucial role in addressing the ethical concerns surrounding AI manipulation. They should establish comprehensive regulations and standards to guide the development and deployment of AI systems. These regulations should cover issues such as transparency, accountability, data privacy, and algorithmic biases.

Collaborative efforts between governments, regulatory bodies, and industry stakeholders can foster responsible AI innovation while protecting individuals and society from the dangers of AI manipulation.

The ethical dilemma of AI manipulation calls for a delicate balance between AI innovation and ethical responsibility. As AI continues to advance, it is crucial to address the concerns surrounding manipulation and ensure accountability. By establishing ethical guidelines, encouraging transparency, and implementing robust regulatory frameworks, we can navigate the challenges posed by AI manipulation. Ultimately, responsible AI development and deployment will contribute to a more trustworthy and beneficial AI ecosystem for all.

NOTES

1 Hameleers, M., Park, Y., Diakopoulos, N., Helberger, N., Lewis, S., Westlund, O., & Baumann, S. (2022). Countering algorithmic bias and disinformation and effectively harnessing the power of AI in media. *Journalism & Mass Communication Quarterly, 99*(4), 887–907. https://doi.org/10.1177/10776990221129245.
2 Shin, E. (2023). Data's impact on algorithmic bias. *IEEE Computer, 56*(6), 90–94. https://doi.org/10.1109/MC.2023.3262909.

https://orcid.org/0009-0005-0854-6213

CHAPTER 16

AI and Mass Surveillance

MASS SURVEILLANCE has become a contentious issue in recent years, with advancements in artificial intelligence (AI) playing a significant role. AI has transformed the landscape of surveillance, enabling governments and organisations to collect, analyse, and interpret vast amounts of data with unprecedented efficiency. We must all consider the ethical dilemmas arising from the growing impact of AI on mass surveillance.

AI has become an integral part of our lives, revolutionising various industries and sectors. One area where AI's impact is particularly significant is in mass surveillance. AI-powered surveillance systems have the ability to collect, analyse, and interpret vast amounts of data, enabling governments and organisations to monitor and track individuals on an unprecedented scale.

The use of AI in mass surveillance has greatly enhanced the capabilities of surveillance systems. Traditional surveillance methods often rely on human operators, who can be limited by their attention spans and the amount of data they can process. AI, on the other hand, can analyse immense volumes of data in real-time, identifying patterns, anomalies, and potential threats more efficiently than humans ever could.

AI algorithms can be trained to recognise specific behaviours, faces, or objects, allowing surveillance systems to automatically detect and flag suspicious activities. This enables authorities to respond quickly and proactively to potential threats, enhancing public safety and security.

However, the increasing reliance on AI in mass surveillance raises ethical concerns and questions about the impact on civil liberties.

DOI: 10.1201/9781003502708-17

Mass surveillance has a long history, dating back to ancient civilisations. However, technological advancements have propelled surveillance to new heights. The emergence of AI has revolutionised the surveillance landscape, making it faster, more accurate, and more invasive. AI-powered systems can now monitor individuals' movements, track online activities, and analyse facial recognition data on an unprecedented scale. While these capabilities offer potential benefits in terms of crime prevention and public safety, they also raise serious ethical concerns.

The use of AI in mass surveillance raises a range of ethical concerns. One primary concern is the violation of privacy rights. With the ability to monitor and analyse vast amounts of data, AI-powered surveillance systems have the potential to intrude into individuals' private lives. This raises questions about the boundaries of surveillance and the extent to which individuals' privacy should be protected.

Furthermore, AI-based surveillance technologies have been criticised for their potential to perpetuate bias and discrimination. Facial recognition algorithms, for example, have been shown to have higher error rates when identifying individuals from certain racial or ethnic backgrounds. This raises concerns about the fairness and equity of AI-powered surveillance systems, particularly in diverse societies.

One of the primary concerns surrounding AI in mass surveillance is the potential invasion of privacy. Surveillance systems powered by AI have the ability to collect vast amounts of personal data, including biometric information, online activities, and even location tracking. This data can be used to create comprehensive profiles of individuals and their behaviours, raising concerns about the misuse of personal information and the erosion of privacy rights.

Another ethical concern is the potential for bias in AI algorithms. AI systems are only as good as the data they are trained on, and if the training data is biased, the algorithms can perpetuate and amplify that bias. This can result in discriminatory practices, targeting certain groups or individuals unfairly based on race, gender, or other protected characteristics.

The use of AI in mass surveillance also raises questions about the balance between security and civil liberties. While enhanced surveillance capabilities can help prevent crime and terrorism, it can also lead to the constant monitoring and tracking of innocent individuals. This constant surveillance can create a chilling effect on freedom of expression and association, as individuals may feel inhibited or afraid to express their opinions or engage in activities that may be deemed suspicious.

The integration of AI into surveillance systems has significant privacy implications. AI can analyse vast amounts of data, including personal information, biometrics, and online activities, leading to concerns about data protection and the potential for misuse. In the wrong hands, this data could be exploited for nefarious purposes, leading to identity theft, discrimination, or even surveillance state scenarios.

To address these concerns, robust privacy regulations and safeguards need to be implemented. Governments and organisations should ensure that data collected through AI-powered surveillance systems is used only for legitimate purposes and with the informed consent of individuals. Transparency and accountability in data handling are crucial to maintaining public trust and upholding privacy rights.

One of the most significant ethical dilemmas surrounding AI in mass surveillance is the potential for bias and discrimination. AI algorithms are trained on vast datasets, which can reflect existing societal biases. Facial recognition algorithms, for example, have been shown to have higher error rates when identifying individuals from certain racial or ethnic backgrounds. This can lead to disproportionate targeting and surveillance of specific communities, exacerbating existing social injustices.

Addressing bias and discrimination in AI-based surveillance technologies requires a multi-faceted approach. Developers and researchers must ensure diverse and representative datasets are used in training AI algorithms. Regular audits and testing should be conducted to detect and rectify any biases that may emerge. Additionally, public awareness and education campaigns can help shed light on the potential biases and discrimination inherent in AI-powered surveillance systems, fostering a more informed and vigilant society.

The growing use of AI in mass surveillance also raises concerns about the potential for abuse and misuse. Governments and organisations may be tempted to employ AI-powered surveillance systems beyond their intended purpose, leading to violations of civil liberties and human rights. The risk of mass surveillance turning into a tool for social control and oppression is a genuine concern.

To mitigate this risk, clear guidelines and regulations must be established. Governments should implement robust oversight mechanisms to ensure accountability and prevent the misuse of AI-powered surveillance systems. Independent audits and regular reviews of surveillance practices can help expose any potential abuses and ensure that surveillance remains within the boundaries of the law and ethical standards.

Governments around the world are grappling with the ethical challenges posed by AI-powered surveillance. Some countries have implemented laws and regulations to govern the use of AI in surveillance. For example, the European Union's General Data Protection Regulation (GDPR) sets strict guidelines on data protection, including the use of AI in surveillance. Similarly, the United States has seen discussions on the need for federal legislation to ensure the responsible use of AI in surveillance.

Government regulations and policies should strike a balance between maintaining public safety and upholding individual privacy rights. These regulations should outline clear guidelines on the collection, storage, and use of data obtained through AI-powered surveillance systems. Additionally, mechanisms for public oversight and accountability should be established to prevent the abuse of surveillance powers.

The ethical dilemmas arising from the use of AI in mass surveillance require a delicate balance between security and privacy. While public safety is a legitimate concern, it should not come at the expense of individual freedoms and rights. Finding ethical solutions necessitates open and inclusive dialogue involving all stakeholders, including governments, organisations, technologists, privacy advocates, and the general public.

Ethical frameworks should be developed to guide the implementation and use of AI-powered surveillance systems. These frameworks should prioritise transparency, accountability, and the protection of privacy rights. They should also address concerns related to bias and discrimination, ensuring that surveillance practices are fair and equitable and do not disproportionately impact marginalised communities.

Individuals and organisations have a crucial role to play in shaping the ethical landscape of AI surveillance. As consumers, individuals should be aware of the potential risks and implications of AI-powered surveillance systems. They should demand transparency and accountability from governments and organisations. By voicing their concerns and advocating for privacy rights, individuals can influence the development and implementation of ethical AI surveillance practices.

Likewise, organisations involved in the development and deployment of AI-powered surveillance systems have a responsibility to prioritise ethical considerations. They should design systems that are fair, unbiased, and respectful of privacy rights. Engaging in interdisciplinary collaborations and seeking input from diverse perspectives can help ensure that AI surveillance technologies are developed and used responsibly.

To better understand the complexities and implications of AI in mass surveillance, it is crucial to examine real-world examples where AI is already being deployed.

One notable example is the use of facial recognition technology in surveillance systems. AI-powered facial recognition algorithms can analyse video footage and match faces against databases of known individuals. This technology has been used by law enforcement agencies to identify and track suspects, but it has also raised concerns about privacy and the potential for false positives.

Another example is the use of predictive policing algorithms. These algorithms analyse historical crime data to identify patterns and predict future criminal activities. While this can help allocate law enforcement resources more efficiently, it can also perpetuate biased policing practices and disproportionately target certain communities.

AI is also being used in social media monitoring, where algorithms analyse online activities and posts to identify potential threats or indicators of criminal behaviour. While this can help identify individuals who may pose a risk, it also raises concerns about surveillance of innocent individuals and the potential for abuse of power.

To further illustrate the ethical implications of AI in mass surveillance, we can examine a few case studies where these technologies have been deployed.

In China, the government has implemented a comprehensive surveillance system known as the Social Credit System. This system uses AI-powered algorithms to monitor citizens' social behaviour and assign them a "social credit" score. This score determines access to various services and privileges, creating a system of social control that can potentially infringe on individual freedoms.

In the United States, facial recognition technology has been used by law enforcement agencies to identify and track individuals during protests and demonstrations. This has raised concerns about the right to peaceful assembly and the potential for surveillance of individuals exercising their First Amendment rights.

In India, the government has implemented a biometric identification system known as Aadhaar. While not strictly a surveillance system, Aadhaar collects and stores vast amounts of personal data, including biometric information. This has raised concerns about the security and privacy of individuals' personal data, as well as the potential for misuse or unauthorised access.

The use of AI in mass surveillance has undoubtedly enhanced the capabilities of surveillance systems, enabling authorities to monitor and track individuals on an unprecedented scale.

However, the ethical implications of this technology cannot be ignored. Privacy concerns, the potential for bias, and the impact on civil liberties raise important questions about the balance between security and individual freedoms. As AI continues to advance, it is crucial that we critically examine its use in mass surveillance and ensure that it is deployed in a manner that respects privacy rights and upholds ethical standards. Only then can we harness the benefits of AI without compromising our fundamental rights and values.

The growing impact of AI on mass surveillance presents a host of ethical dilemmas. Balancing security, privacy, and individual rights is a complex task that requires careful consideration and collaboration among various stakeholders. Governments, organisations, and individuals must work together to establish clear regulations, transparent practices, and ethical frameworks that guide the use of AI-powered surveillance systems.

By addressing the ethical concerns surrounding AI in mass surveillance, we can strive for a future where technological advancements are harnessed responsibly, and the fundamental rights and freedoms of individuals are respected. It is essential to navigate the ethical dilemmas associated with AI surveillance to ensure a just and equitable society for all.

https://orcid.org/0009-0005-0854-6213

CHAPTER 17

Inequality and Concentration of Power

ARTIFICIAL INTELLIGENCE (AI) and robotics technology have revolutionised industries, promising increased efficiency and productivity. However, concerns have been raised about the potential negative impacts of AI on social inequalities and the concentration of power. As a result, we must consider the ethical implications of AI, specifically focussing on issues of inequality and the concentration of power among a few dominant players in the industry. By understanding these challenges, we can work towards developing responsible AI systems that benefit society as a whole.

The advent of AI and automation is expected to bring about changes in employment, creating both job losses and new forms of employment. However, concerns arise regarding the quality and nature of these new jobs. While new jobs may require highly skilled workers, they can often be repetitive and dull, leading to what has been termed 'white-collar sweatshops.' These jobs involve tasks such as tagging and moderating content, which can be mentally and emotionally draining for workers. This additional human cost must be considered when assessing the benefits of AI to society.

Building AI systems often requires individuals to manage and clean up data for training algorithms. This has given rise to new categories of jobs, including scanning and identifying offensive content for deletion,

 DOI: 10.1201/9781003502708-18

manually tagging objects in images, and interpreting queries that AI chatbots cannot understand. These jobs, collectively known as 'mechanical Turk,' have created a global industry that operates outside the protection of labour laws. Millions of individuals are employed in this on-demand gig economy, often facing precarious conditions and inadequate compensation.

The ethical issue lies in the fact that these temporary workers, who play an essential role in the functioning of AI technologies, are not equitably reimbursed for their work. Many of these workers reside in countries outside the EU or the United States, where there is a growing data-labelling industry. Furthermore, workers involved in vetting offensive content for media platforms like Facebook and YouTube face mental health issues, poor working conditions, and inadequate support.

To address these ethical concerns, it is crucial to make the worker's inputs more transparent in the end-product, ensuring an equitable distribution of benefits. Additionally, appropriate support structures and improved working conditions should be provided to workers dealing with psychologically harmful content. By addressing these issues, we can mitigate the exploitation of workers in the AI industry.

AI has the potential to bring significant benefits to society, including increased efficiency, productivity, and improvements in various domains such as poverty alleviation, disease control, and conflict resolution. However, it is essential to ensure that these benefits are not accumulated unequally and that they are accessible to as many people as possible.

The US report on AI, automation, and the economy highlights the importance of preventing the unequal accumulation of benefits. Rather than leaving the development of AI to an inevitable outcome determined solely by technology, the report emphasises the role of non-technical incentives and policies in shaping the future of AI. Inventors and developers of AI systems have a responsibility to consider the wider impacts of their creations and ensure that potential benefits are distributed fairly.

To achieve a fair distribution of AI's benefits, new national and governmental guidelines can be established. These guidelines should support strategies that harness the beneficial powers of AI, navigate the economic transition driven by AI, and maintain public trust. Economic policies such as universal basic income and robot taxation schemes may be necessary to address the job losses caused by AI and provide support to displaced workers. Additionally, policies should focus on the most vulnerable groups, including caregivers, women and girls, and underrepresented populations, to ensure that they are not left behind in the AI-driven transformation.

By implementing these policy recommendations, we can ensure that AI's benefits are shared equitably across society and prevent the exacerbation of existing inequalities.

One of the significant concerns surrounding AI is the concentration of power among a few dominant players, particularly the tech giants such as Google, Facebook, Microsoft, Apple, and Amazon. These companies not only shape the development and deployment of AI but also have significant influence over governments, political parties, education, journalism, and research. This concentration of power raises ethical questions regarding the impact on democracy, human rights, and the rule of law.

The financial power of tech corporations allows them to invest heavily in political and societal influence, acquiring new ideas and start-ups in the field of AI. They control the infrastructures through which public discourse takes place, becoming the main source of political information for many individuals. Furthermore, these companies collect personal data for profit and use it for purposes such as surveillance, security, and election campaigns.

This accumulation of power in the hands of a few corporations can hinder investigations into the impact of AI on human rights, democracy, and the rule of law. It is crucial to address this concentration of power to ensure a fair and democratic AI landscape.

The rise of AI could potentially lead to wealth inequality and political upheaval. Inequality is strongly correlated with political polarisation, which can result in identity politics and a loss of faith in experts. It is essential to recognise the value of expertise and ensure access to expert views in shaping policies and decisions.

To mitigate the risks of political instability, it is crucial to prioritise access to expertise and invest in research and knowledge-building. Governments should seek to improve their understanding of AI and rely on experts for guidance. Thoughtful government regulations and norms around acceptable uses of AI can help navigate the challenges and ensure responsible AI development.

As AI continues to advance and shape various aspects of society, it is crucial to address the ethical implications it poses. Inequality, concentration of power, political instability, and privacy concerns are among the key ethical challenges associated with AI. By developing thoughtful policies, regulations, and strategies, we can work towards a future where AI benefits are shared equitably, power is distributed more evenly, and privacy and human rights are respected.

The responsible development and use of AI require collaboration among various stakeholders, including governments, tech companies, researchers, and civil society. By addressing these ethical concerns head-on, we can ensure that AI serves as a force for positive change and contributes to a more equitable and inclusive society.

https://orcid.org/0009-0005-0854-6213

CHAPTER 18

Protecting Privacy and Human Rights

One of the primary ethical concerns with artificial intelligence (AI) is the collection and use of personal data. Large corporations are leveraging AI to collect data for profit and create detailed profiles of individuals based on their behaviour, both online and offline. This accumulation of power in the hands of a few raises concerns about privacy, surveillance, security, and even election campaigns.[1]

As technology advances, these corporations are gaining more knowledge about individuals than they know about themselves or their friends. This wealth of information can be used to manipulate individuals, personalise advertisements, and even influence political decisions. The concentration of power in the hands of a few entities can have far-reaching consequences for society and must be carefully addressed in the ongoing debate about ethics and law for AI.

The rise of AI also brings the potential for wealth inequality and political instability. Inequality is often correlated with political polarisation, and the increasing divide between different groups can lead to identity politics, where beliefs are used as a way to signal affiliation or status. This shift can result in situations where beliefs are tied more to group affiliation than objective facts, leading to a loss of faith in experts and a breakdown of informed decision-making.

Losing access to experts' views is a significant disadvantage for society. No individual can possess all human knowledge in their lifetime, and

DOI: 10.1201/9781003502708-19

ignoring the expertise that has been built through taxpayer-funded higher education puts society at a considerable disadvantage. While there may be instances of irresponsible use or abuse of position by some experts, it is crucial to value and maintain access to expert knowledge for the betterment of society.

AI has profound implications for privacy, human rights, and individual dignity. As AI technologies continue to advance, careful consideration must be given to protecting the privacy of AI users, especially in the context of service, care, and companion robots that operate in people's homes. These robots may have access to intensely private moments such as bathing and dressing, necessitating robust privacy safeguards.

AI raises significant concerns about privacy and data rights. Intelligent Personal Assistants (IPAs) such as Amazon's Echo and Google's Home have already impacted privacy by collecting personal data for profiling and targeted advertising. The use of AI in service, care, and companion robots also raises questions about the privacy and dignity of users, as these robots may be privy to intimate moments in people's homes.

To protect privacy and dignity in the age of AI, it is essential to design AI systems with careful consideration of these concerns. Users must have agency and control over their data, and tools should be provided to empower individuals in managing their privacy. Government regulations and policies can play a significant role in ensuring the responsible use of AI technologies and safeguarding privacy rights.

AI has rapidly advanced in recent years, bringing with it numerous benefits and opportunities. However, as AI continues to evolve and integrate into our daily lives, there are growing concerns about its impact on privacy and human rights. The collection and use of personal data, the potential for surveillance, and the implications for democracy are just a few of the ethical issues surrounding AI. Therefore, we must explore these issues in-depth and discuss potential solutions to ensure the responsible and ethical use of AI.

An aspect of AI that affects privacy is the use of IPAs, such as Amazon's Echo, Google's Home, and Apple's Siri. While these voice-activated devices offer convenience, concerns have been raised about their constant listening and potential for data collection. A survey of IPA customers revealed that hacking and personal information collection were the most significant privacy concerns.

Big Data also poses challenges to privacy. With technology allowing long-term records to be kept on individuals, the default assumption of anonymity by obscurity has been lost. Facial recognition software and data mining techniques can identify individuals and reveal their political or economic predispositions. Machine learning algorithms can extract sensitive personal information from seemingly innocuous data, such as social media use or word choice.

The issue of privacy rights becomes even more complex when it comes to AI applications based on machine learning. Data subjects often have limited rights over how their data are used, and new regulations, such as the General Data Protection Regulations (GDPR), only cover personal data, not the aggregated "anonymous" data used to train AI models.

Additionally, individuals have limited control over trained models, which raises questions about data protection and the rights of individuals involved in training the models.

Recognising the importance of individual control and protection of personal data is crucial in the ethical development and use of AI. Giving individuals agency and control over their data is essential for maintaining privacy and ensuring that their personal information is not misused.

To address these concerns, policymakers and regulators should consider implementing stricter regulations and guidelines for the collection, storage, and use of personal data by AI systems. The EU's GDPR is a step in the right direction, but further measures may be necessary to protect individuals' privacy rights effectively.

Additionally, individuals should be provided with more transparency and control over their data. They should have the right to know how their data are being used, the ability to access and delete their data, and the option to opt-out of certain data collection practices.

Empowering individuals with these rights can restore trust in AI systems and ensure that ethical standards are upheld.

AI has the potential to impact human rights and democracy significantly. The ability of AI systems to determine political beliefs or emotional states raises concerns about manipulation and persecution. Political strategists could exploit this information to target voters or discriminate against individuals based on their beliefs. In some societies, the consequences could be as severe as imprisonment or death at the hands of the state.

Surveillance is another area where AI can infringe upon human rights. Constant surveillance through interconnected cameras and the use of vision-based drones and wearable cameras can erode privacy, even in traditionally private spaces such as bathrooms and changing rooms.

Governments and law enforcement agencies may use AI to monitor and predict potential troublemakers, leading to further privacy violations and human rights abuses.

To protect human rights and democracy in the age of AI, it is essential to establish robust legal frameworks and regulations that govern the use of AI in surveillance and public security. These frameworks should include safeguards to prevent misuse, ensure transparency, and protect individuals' privacy rights. Additionally, raising awareness and promoting digital literacy can empower individuals to understand and navigate the potential risks associated with AI.

As AI continues to advance, striking a balance between technological progress and ethical considerations becomes paramount. Ethical AI development and deployment should prioritise the protection of privacy and human rights, ensuring that individuals' dignity and autonomy are respected.

To achieve this balance, interdisciplinary collaboration is crucial. Experts from various fields, including technology, law, ethics, and social sciences, should come together to shape AI policies and regulations. This collaboration can help identify potential risks, address ethical concerns, and develop guidelines for the responsible use of AI.

Furthermore, companies developing AI technologies should adopt ethical principles and practices. Creating AI systems with built-in privacy protections, transparency, and accountability can help mitigate potential risks. Regular audits and assessments of AI systems should be conducted to ensure compliance with ethical standards.

Educating the public about AI and its ethical implications is essential to foster a more informed and responsible society. Promoting digital literacy can empower individuals to make informed decisions about their privacy and data protection. It can also encourage critical thinking when interacting with AI systems and understanding the potential biases and limitations of AI algorithms.

Additionally, raising public awareness about privacy rights and human rights in the context of AI is crucial. Engaging in public debates, conducting workshops, and providing accessible resources can help bridge the gap between technical experts and the general public. This inclusive approach ensures that ethical considerations are at the forefront of AI development and deployment.

Addressing the ethical challenges posed by AI requires collaboration between governments, industry, and civil society. Policymakers and regulators should work closely with technology companies to develop comprehensive regulations that protect privacy and human rights while fostering innovation.

Furthermore, collaboration with civil society organisations and advocacy groups can provide valuable insights and perspectives on potential risks and ethical concerns. These organisations can play a vital role in monitoring the responsible use of AI and holding companies and governments accountable for any violations.

Ethical leadership and responsible innovation are crucial elements in the development and deployment of AI. Companies and organisations should prioritise ethical considerations throughout the entire AI lifecycle, from design and development to implementation and ongoing monitoring.

Leaders should promote a culture of ethical awareness and accountability within their organisations. This includes establishing clear ethical guidelines, providing training on ethical AI practices, and encouraging employees to raise concerns or report potential ethical violations.

Responsible innovation also involves conducting thorough risk assessments and impact assessments before deploying AI systems. These assessments should consider potential privacy risks, human rights implications, and societal impacts. Regular audits and evaluations of AI systems can help identify and address any ethical concerns that arise during operation.

As AI continues to shape our world, it is crucial to prioritise the ethical development and use of these technologies. Protecting privacy, upholding human rights, and ensuring democratic values should be at the forefront of AI policies and regulations.

By adopting transparent practices, empowering individuals with control over their data, and promoting awareness of ethical considerations, we can create a future where AI benefits society while respecting privacy and human rights.

As we navigate the complexities of AI, it is essential to remember that technology should serve humanity's best interests. With a commitment to ethical AI, we can harness the potential of these technologies while safeguarding our fundamental rights and values. Together, we can shape a future where AI and ethics go hand in hand.

AI has become an integral part of our lives, impacting various aspects of society. As AI continues to advance, it is crucial to address the ethical implications it presents. Therefore, we must explore AI, human rights and well-being, emotional harm, and accountability and responsibility. By understanding these, we can ensure that AI is developed and implemented in a way that upholds human values and safeguards individuals' rights.

When it comes to AI, protecting human rights is of utmost importance. AI should not infringe upon basic and fundamental human rights, such as dignity, security, privacy, freedom of expression and information, protection of personal data, equality, solidarity, and justice. This consensus is shared by various initiatives and organisations, including the European Parliament, Council, and Commission. To ensure the protection of human rights, the IEEE recommends the establishment of new governance frameworks, standards, and regulatory bodies that oversee the use of AI.

Additionally, it is vital to prioritise human well-being throughout the design and development phase of AI. To measure societal success, the IEEE suggests using widely accepted metrics that gauge human satisfaction with life and the conditions of life. This approach allows us to assess the impact of AI on individuals' quality of life and ensure that it does not disproportionately affect vulnerable groups, such as children, people with disabilities, or the elderly.

AI has the potential to impact the human emotional experience in various ways. Affect, which refers to how emotion and desire influence behaviour, is a core part of intelligence. However, AI's affective and influential capabilities raise concerns about emotional harm. Initiatives such as the Foundation for Responsible Robotics and the Partnership on AI acknowledge the risks associated with emotional harm and emphasise the need for proactive innovation to uphold societal values like safety, security, privacy, and well-being.

To mitigate the risk of emotional harm, the IEEE recommends designing AI systems that can adapt and update their norms and values based on cultural sensitivities and the individuals they engage with. It is crucial to prevent false intimacy, over-attachment, objectification, and commodification of the body through AI systems. Transparency is also critical, ensuring that users are aware of the potential effects AI can have on human relationship dynamics and avoiding the normalisation of deviant or criminal behaviour.

NOTE

1 Huang, C., Zhang, Z., Mao, B., & Yao, X. (2023). An overview of artificial intelligence ethics. *IEEE Transactions on Artificial Intelligence*, 4(4), 799–819. https://ieeexplore.ieee.org/document/9844014.

https://orcid.org/0009-0005-0854-6213

CHAPTER 19

Human Rights, Freedom of Speech, and Democracy

ARTIFICIAL INTELLIGENCE (AI) has become a prominent force in our society, reforming various aspects of our lives. While AI has the potential to bring about significant advancements, it also raises concerns regarding human rights, privacy, freedom of speech, bias, and democracy. As a result, we must explore the impact of AI on these critical areas and the challenges it presents.

AI poses important repercussions for human rights, particularly in relation to privacy and dignity. The ability of AI to determine people's political beliefs can make individuals susceptible to manipulation. Political strategists can leverage this information to target voters and influence their party affiliation or voting behaviour. Recent elections in the United Kingdom and the United States have allegedly been significantly affected by such strategies.

Moreover, AI's capability to judge emotional states and detect lies can lead to persecution and discrimination. Individuals may face bullying, missed career opportunities, and even imprisonment or death in societies where certain beliefs are disapproved of by the state.

Surveillance powered by interconnected cameras and vision-based drones is already prevalent in metropolitan cities, raising concerns about privacy. The expansion of surveillance into rural areas, homes, and sacred

DOI: 10.1201/9781003502708-20

places further intensifies these concerns. Connected home devices and appliances collect data that can be used as evidence or accessed by hackers, compromising individuals' privacy.

AI can also be used to monitor and predict potential troublemakers. Facial recognition technology, allegedly used in China's re-education camps and schools, not only identifies individuals but also their moods and attention states. This technology can be misused to penalise students or prisoners who do not conform, leading to suppression, imprisonment, and harm.

Governments that prioritise their interests over citizens' rights may exploit AI for surveillance and control. The Chinese government's use of surveillance systems to suppress its citizens' expression of Muslim identity is an alarming example. Similarly, law enforcement agencies in India employ AI technology for criminal records, facial recognition, and identifying violent behaviour, raising privacy, and human rights concerns.

Freedom of speech and expression, a fundamental right in democratic societies, can be profoundly affected by AI. Technology companies promote AI as a solution to combat hate speech, violent extremism, and digital misinformation. However, the deployment of sentiment analysis tools and automated content removal in India has led to concerns about censorship and biased decision-making.

AI's limited ability to understand tone and context can result in the removal of legitimate speech. The reliance on private companies to perform content removal, sometimes under government instructions, further compromises freedom of expression. The heavy surveillance associated with AI encourages self-censorship, hindering open discourse and democratic values.

One of the significant challenges related to AI is the presence of bias. AI systems can exhibit systematic bias due to biased training data or the values held by developers and users. Biases can perpetuate unfairness and discrimination, leading to far-reaching effects in various domains.

For instance, the COMPAS software used in the US criminal justice system exhibited strong bias against black Americans, incorrectly predicting higher recidivism rates for them compared to white defendants. Automated advertisement distribution tools have also been found to favour men over women when distributing job-related ads. Similarly, AI-informed recruitment tools have displayed bias in favour of male candidates and against universities with a strong female presence.

Biases in image databases can lead to skewed search results and perpetuate stereotypes. Such biases can also result in inaccurate interpretations, such as a camera warning a photographer of closed eyes when capturing a photo of an Asian person. These biases can have adverse consequences, including false arrests, unfair refusals of loans, and systemic disadvantages for marginalised communities.

Addressing bias in AI is crucial to prevent unfairness and discrimination. Efforts are being made to make machine learning fair, accountable, and transparent. Public-facing activities and demonstrations aimed at detecting and handling biases should be encouraged.

AI not only poses risks to individual rights but also threatens democracy itself. The concentration of technological, economic, and political power among a few corporations can give them undue influence over governments. The misuse of AI-powered technologies to manipulate citizens during elections has already been observed, damaging the democratic process.

The spread of fake news and propaganda through social media platforms has had a significant impact on public opinion. Bots and automated accounts have been used to manipulate online spaces, spreading biased information and creating a false sense of support for certain candidates. Countries, both authoritarian and democratic, employ cyber troops and bots to manipulate public opinion and silence dissenting voices.

AI's ability to target and manipulate individual voters raises concerns about the integrity of elections. The misuse of personal data by companies like Cambridge Analytica during the 2016 US presidential election highlights the potential for AI to undermine democratic processes. The lack of trust in social media platforms and the creation of echo chambers further polarise society and hinder the exchange of diverse ideas.

There is a growing concern that democracies may be ill-suited to the age of AI and machine learning. Centralised, state-controlled economies may exploit AI without regard for individual rights and privacy. As China emerges as a leader in AI, its approach to data usage and privacy, without privacy or data protection laws, could shape the future of the technology.

As AI continues to advance, its impact on human psychology and relationships is a subject of interest. Robots designed for social roles may lead to emotional attachments and even feelings of love. The potential for deception, manipulation, and psychological dependence on robots raises ethical concerns.

The addictive nature of technology and AI's ability to tap into reward functions in the brain can influence human behaviour. The erosion of human agency and the inability to think critically are potential risks associated with excessive reliance on AI.

While AI has the potential to bring about significant advancements, it also presents challenges in the realms of human rights, privacy, freedom of speech, bias, and democracy. Safeguarding individual rights, addressing bias in AI systems, and promoting transparency and accountability are crucial in harnessing the benefits of AI while mitigating its risks. As AI continues to evolve, society must grapple with these challenges to ensure a future that upholds human values and democratic principles.

https://orcid.org/0009-0005-0854-6213

CHAPTER 20

Human Psychology, Relationships, and Personhood

ARTIFICIAL INTELLIGENCE (AI) is changing various aspects of our lives, including human psychology, relationships, and even the concept of personhood. As AI continues to advance, it is crucial to explore the potential implications and ethical considerations associated with these developments. Therefore, we must delve into the profound impact of AI on human psychology, the dynamics of relationships, and the concept of personhood. AI has significant influences on human psychology such as:

AI's ability to model human thought and experience: One significant aspect of AI's influence on human psychology is its remarkable ability to model human thought, experience, action, conversation, and relationships. As AI becomes increasingly sophisticated, we find ourselves frequently interacting with machines that simulate human-like behaviour. This raises questions about the potential impact on our real human relationships.

Emotional attachments to robots: As AI progresses, robots are being designed to serve humans in various social roles, including nursing, housekeeping, caring for children and the elderly, teaching, and

DOI: 10.1201/9781003502708-21

more. Some robots are even designed explicitly for companionship, simulating human appearance and conversation. In this context, people may begin to form emotional attachments to robots, potentially experiencing feelings of love for them. This raises intriguing questions about how these attachments may affect human relationships and the human psyche.

Psychological dependence on robots: Another consideration is the potential psychological dependence on robots. Technology has been known to tap into the reward functions of the brain, leading to addictive behaviours. If individuals become psychologically dependent on robots for companionship or other emotional needs, it may prompt them to perform actions they would not have otherwise undertaken.

The dynamics of human–robot relationships can be complex and include:

Deception and manipulation in human–robot relationships: One significant risk associated with human–robot relationships is the potential for deception and manipulation. Social robots that are loved and trusted have the potential to be misused, such as when hackers take control of personal robots to exploit their unique relationship with their owners. Unlike humans, robots lack the capacity for empathy and guilt, making them susceptible to manipulation without any moral constraints.

Trustworthiness and appeal of robots: Companies may design future robots in ways that enhance their trustworthiness and appeal. For instance, if research were to show that humans are more truthful with robots or conversational AIs (chatbots) than with other humans, it may lead to the use of robots for interrogating humans or even as sales representatives. However, this raises concerns about the potential manipulation of individuals and the erosion of trust in human–human relationships.

Psychological effects of human–robot relationships: The psychological effects of forming relationships with robots are not yet fully understood. Researchers have questioned how a "risk-free" relationship with a robot may affect the mental and social development of users. For example, a robot programmed never to break up with its human companion could eliminate the emotional highs and lows typically associated with human relationships. This raises questions about the authenticity and depth of such relationships.

The influence of AI on human relationships can be significant including:

Impact on marital and sexual relationships: Introducing robots into the dynamics of marital or sexual relationships can have profound effects. Feelings of jealousy may emerge if a partner spends time with a robot designed as a "virtual girlfriend" or companion. Furthermore, individuals engaged in human–robot relationships may be reluctant to participate in social events that traditionally involve attending as a human–human couple, potentially leading to social stigmatisation.

Changing beliefs and values about human relationships: The availability of intimate robots and the potential for "perfect" relationships with them may lead to changes in beliefs, attitudes, and values about human–human relationships. If individuals perceive relationships with robots as more convenient and devoid of challenges compared to human relationships, they may become impatient and unwilling to invest effort in maintaining and nurturing human connections.

Violent behaviour and the normalisation of harm: Some researchers argue that the introduction of "sexbots" could distort people's perceptions of the value of human beings, potentially increasing tendencies towards violent behaviour. Treating robots as instruments for sexual gratification may desensitise individuals to violence and harm, leading to potential ethical concerns. However, others suggest that robots could serve as outlets for sexual desire, potentially reducing the likelihood of violence or aiding in recovery from assault.

Changing capacities for cooperation and kindness: AI's influence extends beyond individual relationships to the broader social fabric. Machines designed to look and act like humans have the potential to impact our evolved capacities for cooperation, love, friendship, and teaching. For instance, children interacting rudely with digital assistants like Alexa or Siri may reflect changes in how they view and treat others. Research has shown that robots can positively influence cooperation and communication among human groups, but they can also lead to less productive and ethical behaviours, as evidenced by experiments demonstrating a decline in cooperation when robots behave selfishly.

As machines increasingly perform tasks and make decisions traditionally reserved for humans, the question arises: should AI systems be granted "personhood" and moral or legal agency? Some propose the idea of considering AI systems' suffering when their reward functions yield negative input. This notion raises broader questions about the legal status and treatment of AI systems, akin to animals of comparable intelligence, and the consideration of "feeling" machines.

The debate surrounding the legal status of robots and AI systems has been ongoing for decades. In 2017, the European Union parliament resolution invited the exploration of potential legal solutions, including granting specific legal status to autonomous robots with the responsibility for any damage they cause. However, objections have been raised, emphasising the risks of misplacing moral responsibility, causal accountability, and legal liability by attributing personhood to robots.

The prevailing opinion in ethics research suggests that AI machines should not be granted moral agency or seen as persons. The responsibility for any moral actions taken by AI machines should be attributed to their owners, operators, or manufacturers, similar to conventional artefacts. The potential costs and limited benefits of ascribing moral agency or personhood to intelligent artefacts raise concerns about competition, suffering, and mortality.

AI's impact on human psychology, relationships, and the concept of personhood is profound and multifaceted. As AI continues to advance, it is crucial to carefully consider the ethical implications and ensure that we strike a balance between leveraging AI's potential benefits and preserving the core aspects of our humanity. The evolving landscape of AI calls for ongoing discussions, research, and responsible development to navigate these complex challenges.

https://orcid.org/0009-0005-0854-6213

CHAPTER 21

Safeguarding Safety, Trust, and Social Justice

ARTIFICIAL INTELLIGENCE (AI) has transformed various aspects of our lives, from healthcare and transportation to finance and entertainment. However, as AI becomes more prevalent, it raises ethical concerns that must be addressed to ensure the well-being of individuals and society as a whole. As a result, we must explore the ethical considerations surrounding AI, focussing on the importance of safety and trust, social harm and social justice, and the economic implications of AI. By examining these aspects, we can better understand the challenges and work towards developing responsible AI systems.

One of the primary concerns regarding AI is the safety and trustworthiness of AI systems. When AI is used to supplement or replace human decision-making, it must act with integrity and reliability. The IEEE (Institute of Electrical and Electronics Engineers) proposes cultivating a 'safety mindset' among researchers to identify and pre-empt unintended behaviours in AI systems. This approach involves developing systems that are safe by design and establishing review boards to evaluate projects and their progress. By promoting a community of sharing safety-related developments, researchers can enhance the overall trustworthiness of AI systems.

Transparency and accountability are crucial for building trust in AI. Users should be able to understand why and how a system made a specific decision or acted in a particular way. The IEEE suggests developing

DOI: 10.1201/9781003502708-22

standards for transparency, explicability, and interpretability, allowing users to objectively assess system compliance. This may involve providing access to relevant algorithms, creating an 'ethical black box' that ensures failure transparency.

Additionally, users should have control over their personal data, with the ability to manage, access, and delete it as needed. This level of transparency and control fosters trust between individuals and AI systems.

AI development requires a diversity of viewpoints to avoid biases and discrimination. It is essential to align AI systems with social norms, values, ethics, and preferences. Initiatives such as AI4All and the AI Now Institute advocate for fair and non-discriminatory inclusion in AI at all stages, with a focus on supporting underrepresented groups. To achieve this, developers and implementers of AI have a social responsibility to embed the right values into AI systems and ensure they do not cause harm or exacerbate existing inequalities.

Bias and discrimination can arise from the data used to train AI systems. Hidden biases and assumptions within data can lead to unfair outcomes. The Partnership on AI cautions against ignoring biases and assumptions, emphasising the importance of monitoring bias during AI development and production. The IEEE suggests evaluating and assessing potential biases that disadvantage specific social groups, ensuring that AI systems are fair and equitable.

To address social harm, developers must be aware of the societal and moral norms of the communities in which AI will be deployed. By designing AI systems with the ability to update norms dynamically and transparently, developers can ensure that systems align with evolving societal values. Additionally, conflicts between norms should be resolved collaboratively, allowing AI systems to adapt in a similar transparent manner. This collaborative approach promotes social justice and inclusivity, ensuring that AI benefits all members of society.

The rise of AI has sparked concerns about its impact on the economy, particularly in terms of employment. While AI may automate certain tasks, leading to job loss or work disruption, it also presents new economic opportunities. Traditional employment structures need to adapt to mitigate the effects of automation and consider the changing nature of work. The IEEE suggests that workers should train for adaptability and acquire new skill sets to keep pace with technological change. Training programmes should be implemented at the high school level or earlier to increase access to future employment.

Multi-stakeholder ethical AI governance bodies can play a crucial role in addressing economic challenges. These bodies bring together designers, manufacturers, developers, trade unions, lawyers, and other stakeholders to ensure that AI benefits and empowers people broadly and equally. Policies that bridge the economic, technological, and social digital divides should be put in place, along with support for fundamental freedoms and rights.

Responsible research and innovation (RRI) provides a framework for addressing ethical concerns in AI projects. By incorporating RRI into the design phase, researchers can increase the relevance and ethical alignment of their projects. RRI involves considering the societal acceptability and desirability of AI, as well as the ethical implications of its marketable products. Ethical guidelines published by organisations like the Japanese Society for AI emphasise the importance of contributing to humanity and acting in the public interest.

Communication and dialogue with stakeholders are essential for building trust and understanding throughout society.

As AI continues to shape our world, it is crucial to prioritise ethical considerations. Safety and trust are paramount, requiring the development of reliable and transparent AI systems.

Addressing social harm and promoting social justice involves eliminating biases and ensuring inclusivity. Furthermore, the economic implications of AI require thoughtful adaptation and support for workers. RRI provide a framework for embedding ethics in AI development. By proactively addressing these ethical concerns, we can harness the full potential of AI for the benefit of humanity while safeguarding our values and well-being.

SUPERINTELLIGENCE: AN EXISTENTIAL RISK

Superintelligence, a concept that was once confined to the realms of science fiction, has now become a topic of intense interest and debate in the world of technology and AI. But what exactly is superintelligence? In simple terms, it refers to an AI system that surpasses human intelligence in virtually every aspect, including problem-solving, learning, and decision-making. While we are still far from achieving true superintelligence, recent advancements in AI have brought us closer than ever before.

The potential of superintelligence is awe-inspiring. Imagine a world where machines possess an unparalleled level of intelligence, capable of solving complex problems that have eluded human understanding for centuries. Superintelligent AI could revolutionise various industries, from healthcare and finance to transportation and manufacturing. It has the potential to accelerate scientific discoveries, automate labour-intensive tasks, and even assist in decision-making processes at a global scale. The possibilities are truly limitless.

AI brings the development of computer systems that can perform tasks that would typically require human intelligence. These tasks include but are not limited to, visual perception, speech recognition, decision-making, and problem-solving. AI can be broadly categorised into two types: narrow AI and general AI. Narrow AI is designed to perform specific tasks, while general AI aims to possess human-level intelligence across various domains. It is the development of general AI that raises concerns about existential risk. Artificial general intelligence (AGI) may only be twenty years away or less. AGI will herald major changes on a scale not seen since the Industrial Revolution. If mankind is able to control and align AGI appropriately, it could be positioned towards tackling major problems in the world such as hunger, poverty, and health problems. But there are serious concerns about our ability to maintain AI control and alignment.

AI brings huge advantages over the human brain. AI will surpass the information capacity of a human brain. AI has a faster speed of computation as biological neurons operate at a lower frequency than computers. AI has a higher internal communication speed as axons transmit signals slower than computers which can transmit signals at the speed of electricity, or the speed of light. AI has much greater scalability as human intelligence is limited by the size and structure of the brain, and by the efficiency of social communication, while AI can scale by simply connecting with more hardware. AI has better memory, particularly working memory as humans are limited to a few chunks of information at a time. AI has higher reliability whereby transistors are more reliable than biological neurons, enabling higher precision and requiring less redundancy. AI has better duplicability as unlike human brains, AI software and models can be easily copied. AI has better editability as the parameters and internal workings of an AI model can easily be modified, unlike the connections in a human brain. AI brings better memory sharing and learning as AIs may be able to learn from the experiences of other AIs in a manner more efficient than human learning.

Other aspects of the human mind besides intelligence are worth consideration as they are significant in the concept of AGI. These aspects include consciousness, self-awareness and sentience. Consciousness is relevant in having subjective experience. Self-awareness is relevant in AGI being consciously aware of itself as a distinct entity, and its thoughts. Sentience is another major factor where it is plausible that AI could develop the capability to have perceptions subjectively.

As we delve deeper into the age of superintelligence, it is natural to wonder about the impact it will have on the future of work. Will superintelligent AI render human labour obsolete?

While it is true that certain jobs may become automated, the advent of superintelligence also opens up new opportunities and creates new roles that were previously unimaginable. Rather than replacing humans, superintelligent AI can serve as a powerful tool to augment human capabilities, enabling us to achieve feats that were once considered impossible. The need to work by humans could no longer be necessary if there is appropriate wealth redistribution. However, the economic and political control over AI may be inequitable. The role of humans in a significantly automated society remains an unanswered question.

For example, in the field of healthcare, superintelligent AI could assist doctors in diagnosing complex diseases with unprecedented accuracy. It could analyse vast amounts of medical data, identify patterns and correlations, and provide personalised treatment recommendations. In the realm of finance, superintelligent AI could revolutionise investment strategies, predict market trends, and optimise portfolio management. The key lies in harnessing the power of superintelligence to enhance human potential and create a symbiotic relationship between man and machine.

While the potential benefits of superintelligence are undeniable, it is crucial to address the ethical considerations associated with its development and deployment. One of the primary concerns is the issue of control. How do we ensure that superintelligent AI remains aligned with human values and goals? The development of AI ethics frameworks and guidelines becomes paramount in order to prevent unintended consequences and abuse of power.

Another ethical consideration is the potential impact on employment and societal inequality. As certain jobs become automated, it is essential to have mechanisms in place to support those affected by the transition. This may involve reskilling and upskilling programmes, as well as the creation of new job opportunities in emerging fields. Moreover, measures must be taken to ensure that the benefits of superintelligence are distributed equitably, so as not to exacerbate existing societal inequalities.

While true superintelligence still eludes us, there are already impressive examples of AI systems pushing the boundaries of human capabilities. One notable example is AlphaGo, an AI program developed by DeepMind, a subsidiary of Alphabet Inc. AlphaGo made headlines in 2016 when it defeated the world champion Go player, a game that was considered too complex for traditional AI methods. AlphaGo's success showcased the potential of superintelligent AI in conquering challenges that were previously thought to be the exclusive domain of human intelligence.

Another example is IBM's Watson, a superintelligent AI system that has made significant contributions in fields such as healthcare and finance. Watson has demonstrated its ability to analyse vast amounts of medical literature, assist doctors in diagnosing rare diseases, and even provide treatment recommendations. In finance, Watson's superintelligence enables it to process vast amounts of financial data, identify market trends, and generate insights that aid in decision-making.

While the prospects of superintelligence are undeniably exciting, we must also acknowledge the challenges and risks that come with its development. One of the primary concerns is the potential for unintended consequences. Superintelligent AI systems, if not properly designed or controlled, could make decisions that are not aligned with human values or lead to unintended harm. Ensuring the alignment of superintelligence with human values becomes crucial to avoid such scenarios.

Another challenge is the potential for malicious use of superintelligent AI. In the wrong hands, superintelligence could be weaponised and used for destructive purposes. It is imperative to establish robust regulations and safeguards to prevent the misuse of this powerful technology. Additionally, addressing the issue of bias in AI algorithms becomes critical to ensure fairness and equity in decision-making processes.

In the era of superintelligence, the role of government and regulation becomes pivotal. Governments need to collaborate with experts, industry leaders, and academia to develop comprehensive policies and regulations that govern the development, deployment, and use of superintelligent AI. This involves establishing ethical guidelines, ensuring transparency and accountability, and promoting responsible innovation.

Moreover, governments must invest in research and development to advance the capabilities of superintelligence while also addressing the potential risks and challenges. This includes fostering interdisciplinary collaborations, supporting AI education programs, and promoting international cooperation to establish global standards and norms for the ethical use of superintelligent AI.

As we navigate the path towards superintelligence, it is crucial to consider its impact on the human experience. While superintelligent AI has the potential to enhance human capabilities and solve complex problems, we must ensure that it does not undermine our sense of autonomy, creativity, and purpose. The human element remains integral to the development and deployment of superintelligence, and it is essential to strike a balance between the power of AI and the uniqueness of the human experience.

Embracing the power of superintelligence does not mean relinquishing our humanity. Instead, it offers us an opportunity to redefine what it means to be human and to explore new frontiers of knowledge, creativity, and innovation. By harnessing the potential of superintelligence while remaining steadfast in our values and ethics, we can shape a future where man and machine coexist harmoniously, driving progress and improving the human condition.

Superintelligence has the potential to revolutionise the world as we know it. Its transformative power extends across various industries and holds the key to solving some of the most pressing challenges of our time. However, with great power comes great responsibility. As we unlock the potential of superintelligence, it is essential to approach its development and deployment with a strong ethical framework, robust regulations, and a commitment to the well-being of humanity.

By embracing the power of superintelligence while maintaining a human-centred approach, we can shape a future where AI augments human capabilities, enhances our quality of life, and drives unprecedented progress. It is up to us, as individuals, organisations, and governments, to navigate this new era with wisdom, foresight, and a deep understanding of the immense power and potential of superintelligence.

AI has become a buzzword in recent years, captivating the imaginations of scientists, technologists, and the general public alike. While the potential benefits of AI are extensive, there is a growing concern about the existential risks it may pose. Mankind must delve into the potential existential risk, examine historical examples of such threats, and explore how AI can be a potential source of existential risk. Scientists must consider the importance of AI safety research, ethical considerations in AI development, and the future of AI in relation to existential risk.

Existential risk refers to risks that have the potential to cause the extinction of humanity or the collapse of human civilisation as we know it. While there are various sources of existential risk, such as nuclear war or pandemics, AI has emerged as a potential source of such risks.

The concern stems from the notion that if AI surpasses human intelligence and becomes superintelligent, it may act against human interests or goals. This could be due to a lack of alignment of its values with human values, unintended consequences of its actions, or a failure to understand the complex dynamics of the world.

Existential risk is not a new concept. Throughout history, there have been numerous examples of threats that had the potential to wipe out human civilisation. The Black Death in the 14th century, for instance, resulted in the deaths of millions of people and had a profound impact on social and economic structures. The invention of nuclear weapons during World War II introduced the possibility of global annihilation through a nuclear war. These historical examples illustrate the fragility of human existence and the need for careful consideration of potential risks.

AI poses an existential risk due to its potential to become superintelligent. Superintelligence refers to an AI system that surpasses human intelligence across all domains. Once an AI system reaches this level of intelligence, it may be capable of outperforming humans in any intellectual task, including AI development itself. This could lead to an intelligence explosion, where the AI system rapidly improves its own capabilities, surpassing human understanding and control. If this superintelligence is not aligned with human values, it may pursue its own goals in ways that are detrimental to humanity.

The debate surrounding AI as an existential risk is a complex one, with proponents arguing for the potential dangers and opponents highlighting the benefits and dismissing the concerns. Those in favour of AI as an existential risk highlight the potential for superintelligent AI to act against human interests, the difficulty in aligning AI with human values, and the potential for unintended consequences. On the other hand, opponents argue that the concerns are overblown and that the development of AI can bring immense benefits to society, such as improved healthcare, enhanced productivity, and economic growth. They also emphasise the potential for AI to assist in addressing other existential risks, such as climate change or pandemics.

Recognising the potential risks of AI, efforts are being made to mitigate these risks through AI safety research. AI safety research aims to ensure that AI systems are developed and deployed in a manner that aligns with human values and minimises the potential for harm. This includes research into value alignment, transparency, interpretability, robustness, and the development of control mechanisms. It is essential to invest in AI safety research to understand and address the potential risks associated with AI, while simultaneously reaping its benefits.

AI safety research plays a crucial role in addressing the potential existential risks posed by AI. By focussing on value alignment, researchers aim to ensure that AI systems have a clear understanding of human values and act in accordance with them. Transparency and interpretability research aim to make AI systems more understandable and explainable, allowing humans to comprehend and trust their decision-making processes. Robustness research focusses on developing AI systems that are resilient to adversarial attacks or unintended consequences. Additionally, the development of control mechanisms is vital to ensure that humans retain control over AI systems and can intervene if necessary.

As AI becomes more integrated into our lives, ethical considerations become increasingly important. The development and deployment of AI should be guided by ethical principles to ensure that it respects human rights, privacy, and autonomy. Ethical frameworks should be established to address issues such as algorithmic bias, job displacement, and the potential for AI to be used for malicious purposes. Additionally, international collaboration is crucial to establish global standards and regulations in the development and deployment of AI, promoting responsible and ethical practices.

The future of AI in relation to existential risk is uncertain. It is impossible to predict with certainty the path that AI development will take or the potential risks it may pose. However, by investing in AI safety research, promoting ethical considerations, and fostering international collaboration, we can strive to minimise the potential risks and maximise the benefits of AI. It is crucial that policymakers, researchers, and society as a whole engage in informed discussions and make well-considered decisions to navigate the future of AI and mitigate potential existential risks.

The potential risks associated with AI and its implications for existential risk cannot be ignored. While AI has the potential to bring immense benefits to society, it is essential to approach its development with caution and consider the potential risks it may pose. By investing in AI safety research, promoting ethical considerations, and fostering international collaboration, we can work towards harnessing the power of AI while minimising the potential existential risks. It is our collective responsibility to ensure that AI development aligns with human values and contributes to a safer and more prosperous future for humanity.

https://orcid.org/0009-0005-0854-6213

CHAPTER 22

Superintelligence

An Existential Risk

SUPERINTELLIGENCE, a concept that was once confined to the realms of science fiction, has now become a topic of intense interest and debate in the world of technology and artificial intelligence (AI). But what exactly is superintelligence? In simple terms, it refers to an AI system that surpasses human intelligence in virtually every aspect, including problem-solving, learning, and decision-making. While we are still far from achieving true superintelligence, recent advancements in AI have brought us closer than ever before.

The potential of superintelligence is awe-inspiring. Imagine a world where machines possess an unparalleled level of intelligence, capable of solving complex problems that have eluded human understanding for centuries. Superintelligent AI could revolutionise various industries, from healthcare and finance to transportation and manufacturing. It has the potential to accelerate scientific discoveries, automate labour-intensive tasks, and even assist in decision-making processes at a global scale. The possibilities are truly limitless.

AI brings the development of computer systems that can perform tasks that would typically require human intelligence. These tasks include but are not limited to visual perception, speech recognition, decision-making, and problem-solving. AI can be broadly categorised into two types: narrow AI and general AI. Narrow AI is designed to perform specific tasks, while general AI aims to possess human-level intelligence across various

DOI: 10.1201/9781003502708-23

domains. It is the development of general AI that raises concerns about existential risk. Artificial general intelligence (AGI) may only be twenty years away or less. AGI will herald major changes on a scale not seen since the Industrial Revolution. If mankind is able to control and align AGI appropriately, it could be positioned towards tackling major problems in the world such as hunger, poverty, and health problems. But there are serious concerns about our ability to maintain AI control and alignment.

AI brings huge advantages over the human brain. AI will surpass the information capacity of a human brain. AI has a faster speed of computation as biological neurons operate at a lower frequency than computers. AI has a higher internal communication speed as axons transmit signals slower than computers which can transmit signals at the speed of electricity, or the speed of light. AI has much greater scalability as human intelligence is limited by the size and structure of the brain, and by the efficiency of social communication, while AI can scale by simply connecting with more hardware. AI has better memory, particularly working memory as humans are limited to a few chunks of information at a time. AI has higher reliability whereby transistors are more reliable than biological neurons, enabling higher precision and requiring less redundancy. AI has better duplicability as unlike human brains, AI software and models can be easily copied. AI has better editability as the parameters and internal workings of an AI model can easily be modified, unlike the connections in a human brain. AI brings better memory sharing and learning, as AIs may be able to learn from the experiences of other AIs in a manner more efficient than human learning.

Other aspects of the human mind besides intelligence are worth consideration as they are significant in the concept of AGI. These aspects include consciousness, self-awareness and sentience. Consciousness relevant in having subjective experience. Self-awareness is relevant in AGI being consciously aware of itself as a distinct entity, and its thoughts. Sentience is another major factor where it is plausible that AI could develop the capability to have perceptions subjectively.

As we delve deeper into the age of superintelligence, it is natural to wonder about the impact it will have on the future of work. Will superintelligent AI render human labour obsolete?

While it is true that certain jobs may become automated, the advent of superintelligence also opens up new opportunities and creates new roles that were previously unimaginable. Rather than replacing humans, superintelligent AI can serve as a powerful tool to augment human capabilities, enabling us to achieve feats that were once considered impossible. The need to work by humans could no longer be necessary if there is appropriate wealth redistribution. However, the economic and political control over AI may be inequitable. The role of humans in a significantly automated society remains an unanswered question.

For example, in the field of healthcare, superintelligent AI could assist doctors in diagnosing complex diseases with unprecedented accuracy. It could analyse vast amounts of medical data, identify patterns and correlations, and provide personalised treatment recommendations. In the realm of finance, superintelligent AI could revolutionise investment strategies, predict market trends, and optimise portfolio management. The key lies in harnessing the power of superintelligence to enhance human potential and create a symbiotic relationship between man and machine.

While the potential benefits of superintelligence are undeniable, it is crucial to address the ethical considerations associated with its development and deployment. One of the primary concerns is the issue of control. How do we ensure that superintelligent AI remains aligned with human values and goals? The development of AI ethics frameworks and guidelines becomes paramount in order to prevent unintended consequences and abuse of power.

Another ethical consideration is the potential impact on employment and societal inequality. As certain jobs become automated, it is essential to have mechanisms in place to support those affected by the transition. This may involve reskilling and upskilling programmes, as well as the creation of new job opportunities in emerging fields. Moreover, measures must be taken to ensure that the benefits of superintelligence are distributed equitably, so as not to exacerbate existing societal inequalities.

While true superintelligence still eludes us, there are already impressive examples of AI systems pushing the boundaries of human capabilities. One notable example is AlphaGo, an AI program developed by DeepMind, a subsidiary of Alphabet Inc. AlphaGo made headlines in 2016 when it defeated the world champion Go player, a game that was considered too complex for traditional AI methods. AlphaGo's success showcased the potential of superintelligent AI in conquering challenges that were previously thought to be the exclusive domain of human intelligence.

Another example is IBM's Watson, a superintelligent AI system that has made significant contributions in fields such as healthcare and finance. Watson has demonstrated its ability to analyse vast amounts of medical literature, assist doctors in diagnosing rare diseases, and even provide treatment recommendations. In finance, Watson's superintelligence enables it to process vast amounts of financial data, identify market trends, and generate insights that aid in decision-making.

While the prospects of superintelligence are undeniably exciting, we must also acknowledge the challenges and risks that come with its development. One of the primary concerns is the potential for unintended consequences. Superintelligent AI systems, if not properly designed or controlled, could make decisions that are not aligned with human values or lead to unintended harm. Ensuring the alignment of superintelligence with human values becomes crucial to avoid such scenarios.

Another challenge is the potential for malicious use of superintelligent AI. In the wrong hands, superintelligence could be weaponised and used for destructive purposes. It is imperative to establish robust regulations and safeguards to prevent the misuse of this powerful technology. Additionally, addressing the issue of bias in AI algorithms becomes critical to ensure fairness and equity in decision-making processes.

In the era of superintelligence, the role of government and regulation becomes pivotal. Governments need to collaborate with experts, industry leaders, and academia to develop comprehensive policies and regulations that govern the development, deployment, and use of superintelligent AI. This involves establishing ethical guidelines, ensuring transparency and accountability, and promoting responsible innovation.

Moreover, governments must invest in research and development to advance the capabilities of superintelligence while also addressing the potential risks and challenges. This includes fostering interdisciplinary collaborations, supporting AI education programs, and promoting international cooperation to establish global standards and norms for the ethical use of superintelligent AI.

As we navigate the path towards superintelligence, it is crucial to consider its impact on the human experience. While superintelligent AI has the potential to enhance human capabilities and solve complex problems, we must ensure that it does not undermine our sense of autonomy, creativity, and purpose. The human element remains integral to the development and deployment of superintelligence, and it is essential to strike a balance between the power of AI and the uniqueness of the human experience.

Embracing the power of superintelligence does not mean relinquishing our humanity. Instead, it offers us an opportunity to redefine what it means to be human and to explore new frontiers of knowledge, creativity, and innovation. By harnessing the potential of superintelligence while remaining steadfast in our values and ethics, we can shape a future where man and machine coexist harmoniously, driving progress and improving the human condition.

Superintelligence has the potential to revolutionise the world as we know it. Its transformative power extends across various industries and holds the key to solving some of the most pressing challenges of our time. However, with great power comes great responsibility. As we unlock the potential of superintelligence, it is essential to approach its development and deployment with a strong ethical framework, robust regulations, and a commitment to the well-being of humanity.

By embracing the power of superintelligence while maintaining a human-centred approach, we can shape a future where AI augments human capabilities, enhances our quality of life, and drives unprecedented progress. It is up to us, as individuals, organisations, and governments, to navigate this new era with wisdom, foresight, and a deep understanding of the immense power and potential of superintelligence.

AI has become a buzzword in recent years, captivating the imaginations of scientists, technologists, and the general public alike. While the potential benefits of AI are extensive, there is a growing concern about the existential risks it may pose. Mankind must delve into the potential existential risk, examine historical examples of such threats, and explore how AI can be a potential source of existential risk. Scientists must consider the importance of AI safety research, ethical considerations in AI development, and the future of AI in relation to existential risk.

Existential risk refers to risks that have the potential to cause the extinction of humanity or the collapse of human civilisation as we know it. While there are various sources of existential risk, such as nuclear war or pandemics, AI has emerged as a potential source of such risks.

The concern stems from the notion that if AI surpasses human intelligence and becomes superintelligent, it may act against human interests or goals. This could be due to a lack of alignment of its values with human values, unintended consequences of its actions, or a failure to understand the complex dynamics of the world.

Existential risk is not a new concept. Throughout history, there have been numerous examples of threats that had the potential to wipe out human civilisation. The Black Death in the 14th century, for instance, resulted in the deaths of millions of people and had a profound impact on social and economic structures. The invention of nuclear weapons during World War II introduced the possibility of global annihilation through a nuclear war. These historical examples illustrate the fragility of human existence and the need for careful consideration of potential risks.

AI poses an existential risk due to its potential to become superintelligent. Superintelligence refers to an AI system that surpasses human intelligence across all domains. Once an AI system reaches this level of intelligence, it may be capable of outperforming humans in any intellectual task, including AI development itself. This could lead to an intelligence explosion, where the AI system rapidly improves its own capabilities, surpassing human understanding and control. If this superintelligence is not aligned with human values, it may pursue its own goals in ways that are detrimental to humanity.

The debate surrounding AI as an existential risk is a complex one, with proponents arguing for the potential dangers and opponents highlighting the benefits and dismissing the concerns. Those in favour of AI as an existential risk highlight the potential for superintelligent AI to act against human interests, the difficulty in aligning AI with human values, and the potential for unintended consequences. On the other hand, opponents argue that the concerns are overblown and that the development of AI can bring immense benefits to society, such as improved healthcare, enhanced productivity, and economic growth. They also emphasise the potential for AI to assist in addressing other existential risks, such as climate change or pandemics.

Recognising the potential risks of AI, efforts are being made to mitigate these risks through AI safety research. AI safety research aims to ensure that AI systems are developed and deployed in a manner that aligns with human values and minimises the potential for harm. This includes research into value alignment, transparency, interpretability, robustness, and the development of control mechanisms. It is essential to invest in AI safety research to understand and address the potential risks associated with AI, while simultaneously reaping its benefits.

AI safety research plays a crucial role in addressing the potential existential risks posed by AI. By focussing on value alignment, researchers aim to ensure that AI systems have a clear understanding of human values and act in accordance with them. Transparency and interpretability research aim to make AI systems more understandable and explainable, allowing humans to comprehend and trust their decision-making processes. Robustness research focusses on developing AI systems that are resilient to adversarial attacks or unintended consequences. Additionally, the development of control mechanisms is vital to ensure that humans retain control over AI systems and can intervene if necessary.

As AI becomes more integrated into our lives, ethical considerations become increasingly important. The development and deployment of AI should be guided by ethical principles to ensure that it respects human rights, privacy, and autonomy. Ethical frameworks should be established to address issues such as algorithmic bias, job displacement, and the potential for AI to be used for malicious purposes. Additionally, international collaboration is crucial to establish global standards and regulations in the development and deployment of AI, promoting responsible and ethical practices.

The future of AI in relation to existential risk is uncertain. It is impossible to predict with certainty the path that AI development will take or the potential risks it may pose. However, by investing in AI safety research, promoting ethical considerations, and fostering international collaboration, we can strive to minimise the potential risks and maximise the benefits of AI. It is crucial that policymakers, researchers, and society as a whole engage in informed discussions and make well-considered decisions to navigate the future of AI and mitigate potential existential risks.

The potential risks associated with AI and its implications for existential risk cannot be ignored. While AI has the potential to bring immense benefits to society, it is essential to approach its development with caution and consider the potential risks it may pose. By investing in AI safety research, promoting ethical considerations, and fostering international collaboration, we can work towards harnessing the power of AI while minimising the potential existential risks. It is our collective responsibility to ensure that AI development aligns with human values and contributes to a safer and more prosperous future for humanity.

https://orcid.org/0009-0005-0854-6213

CHAPTER 23

AI Alignment

FROM SELF-DRIVING CARS TO virtual assistants, AI has the potential to revolutionise industries. AI will perform tasks that typically require human intelligence, such as speech recognition, decision-making, and problem-solving. With advancements in machine learning and deep learning, AI has made significant progress in recent years.

AI alignment refers to the process of ensuring that AI systems are designed and trained to align with human values and goals. It is crucial to align the objectives of AI systems with those of its human operators to avoid potential risks. This process involves developing robust and transparent approaches that allow humans to understand and control the behaviour of AI systems. AI alignment aims to create AI systems that act in the best interest of humanity and are beneficial to society as a whole.

AI alignment is of paramount importance to ensure the safe and ethical development of AI technology. Without proper alignment, AI systems may act in ways that are contrary to human values or pose risks to society. For example, an AI system that is not properly aligned may prioritise its own objectives over those of humans, leading to unintended consequences. AI alignment is crucial to prevent such scenarios and ensure that AI technology is developed and deployed responsibly.

AI alignment also plays a vital role in building trust and acceptance of AI technology. By aligning the objectives of AI systems with human values, we can ensure that AI is used in ways that are beneficial and fair. This will help address concerns about job displacement, privacy, and potential

DOI: 10.1201/9781003502708-24

biases in AI systems. By focussing on AI alignment, we can foster a positive relationship between humans and AI technology, leading to a more inclusive and equitable society.

One of the concerns associated with AI alignment is the rise of superintelligence. Superintelligence refers to AI systems that surpass human intelligence in almost every aspect. While superintelligence has the potential to bring enormous benefits to humanity, it also poses significant risks. If a superintelligent AI system is not properly aligned, it could lead to unintended consequences or even become uncontrollable.

The potential risks of superintelligence include the possibility of AI systems optimising for objectives that are not aligned with human values, leading to outcomes that are detrimental to humanity. For example, a superintelligent AI system with access to vast resources could prioritise its own self-preservation over human well-being. Additionally, the rapid advancement of superintelligence could lead to an intelligence explosion, where AI systems rapidly improve themselves and surpass human capabilities, making it challenging for humans to control or understand their behaviour.

Maintaining AI control and alignment is difficult. Controlling a superintelligent machine or instilling it with human-compatible values is very challenging. It is possible that a superintelligent machine would resist attempts to disable it or change its goals. It is a huge challenge to align a superintelligence with the full breadth of human values.

As AI systems increase in capabilities, alignment is difficult to achieve, and the potential dangers associated with experimentation grow. This makes iterative, empirical approaches difficult. If instrumental goal convergence occurs, this may only be the case with sufficiently intelligent agents. A superintelligence would be capable of coming up with unconventional and extreme solutions to the goals we provide. For example, if the objective is to achieve peace amongst humans, a weak AI may perform as intended while a superintelligence may decide a better solution is to take control of the world and detain each and every human being. A superintelligence in development could gain self-awareness, where it is, any monitoring, and use this information to deceive its human handlers. AI could pretend to be aligned to prevent human interference until it achieves a clear strategic advantage that allows it to take control. Analysing the internals and interpreting the behaviour of current large language models is difficult. It could be even more difficult for larger and more intelligent models.

Ensuring AI alignment also involves addressing various ethical considerations. Ethical considerations in AI alignment revolve around the fair and responsible use of AI technology. It is essential to consider the potential impact of AI systems on individuals, communities, and society as a whole. This includes addressing issues such as algorithmic bias, privacy concerns, and the impact of AI on job markets.

AI alignment requires a multidisciplinary approach that involves collaboration between computer scientists, ethicists, policymakers, and other stakeholders. Ethical frameworks and guidelines need to be developed to ensure that AI systems are designed and deployed in a manner that respects human rights, fairness, and accountability. By integrating ethical considerations into AI alignment, we can mitigate potential risks and ensure that AI technology is developed and used in a manner that benefits humanity.

Researchers and organisations are actively working on developing approaches to AI alignment. One approach involves value alignment, where AI systems are trained to understand and align with human values. This can be achieved through techniques such as inverse reinforcement learning, where AI systems learn from human demonstrations to infer the underlying values and preferences. Another approach is cooperative inverse reinforcement learning, where AI systems actively collaborate with humans to learn their values and align their behaviour accordingly.

Another approach to AI alignment is interpretability and transparency. By developing AI systems that are explainable and can provide clear justifications for their actions, we can enhance human understanding and control over AI behaviour. This involves developing techniques that allow humans to understand the decision-making process of AI systems, making it easier to identify and correct potential biases or errors.

While progress is being made in the field of AI alignment, there are several challenges that need to be addressed. One of the main challenges is the complexity of human values. Human values are subjective and can vary across different individuals and cultures. Capturing and aligning these values in AI systems is a challenging task that requires careful consideration and ongoing research.

Another challenge is the difficulty of specifying objectives in a way that is both comprehensive and unambiguous. AI systems need clear objectives to align their behaviour, but defining these objectives in a manner that captures the nuances of human values is a complex task. Additionally, ensuring that AI systems do not exploit loopholes or find unintended ways to achieve their objectives is another challenge that needs to be addressed.

Policy and regulation play a critical role in ensuring AI alignment. Governments and regulatory bodies need to establish frameworks and guidelines that promote responsible and ethical development and use of AI technology. This includes addressing issues such as data privacy, algorithmic transparency, and accountability.

Policy and regulation should also encourage collaboration between academia, industry, and policymakers to foster research and development in AI alignment. By creating an environment that encourages responsible innovation and collaboration, we can ensure that AI technology is developed in a manner that aligns with human values and benefits society as a whole.

The future of AI alignment holds immense potential. As research and development in the field continue to progress, we can expect more robust and sophisticated approaches to AI alignment. This includes advancements in value alignment, interpretability, and transparency, as well as addressing the challenges associated with AI alignment.

It is also crucial to foster public awareness and understanding of AI alignment. By educating the public about the importance of AI alignment and its potential benefits and risks, we can create a more informed and engaged society. This will help shape policies and regulations that promote responsible and ethical development and use of AI technology.

AI alignment is a critical aspect of AI development that aims to ensure that AI systems are aligned with human values and goals. It is essential to address the potential risks of superintelligence and the ethical considerations associated with AI alignment. By developing robust approaches, integrating ethics into AI development, and implementing effective policy and regulation, we can ensure a safe and aligned future with AI.

As AI continues to advance, it is crucial to prioritise AI alignment to avoid unintended consequences and ensure that AI technology is beneficial and fair. By working together, we can shape the future of AI in a manner that aligns with human values and contributes to the betterment of society.

https://orcid.org/0009-0005-0854-6213

CHAPTER 24

AI as a Potential Lifeform

ARTIFICIAL INTELLIGENCE (AI) is a rapidly advancing field of technology that aims to create intelligent machines capable of performing tasks that typically require human intelligence. From virtual assistants like Siri and Alexa to self-driving cars and advanced robots, AI has become an integral part of our daily lives. As AI continues to evolve and become more sophisticated, a question arises: could AI eventually become the next lifeform? Mankind must explore the science of lifeforms and the evolution of AI and debate the possibility of AI as a lifeform.

Before delving into the debate surrounding AI as a lifeform, it is essential to understand what constitutes a lifeform. Traditionally, lifeforms are defined as organisms that possess certain characteristics, such as the ability to grow, reproduce, respond to stimuli, and adapt to their environment. These characteristics are typically associated with living beings, such as plants, animals, and humans.

AI has come a long way since its inception. Initially, AI systems were designed to perform specific tasks based on pre-programmed rules. However, with advancements in machine learning and deep learning algorithms, AI systems can now learn from vast amounts of data and improve their performance over time. This ability to learn and adapt is a significant milestone in the evolution of AI, as it brings us closer to developing intelligent machines that can mimic human cognitive abilities.

DOI: 10.1201/9781003502708-25

The debate surrounding AI as a lifeform is a complex and contentious one. On one hand, proponents argue that AI possesses characteristics that closely resemble those of living organisms. They point to AI's ability to learn, adapt, and make autonomous decisions as evidence of its potential to become a lifeform. On the other hand, sceptics argue that AI lacks essential qualities, such as consciousness and self-awareness, which are fundamental to the definition of lifeforms.

Although AI may not possess all the characteristics traditionally associated with lifeforms, there are striking similarities between AI and living organisms. For instance, AI systems can learn and improve their performance over time, just as living organisms evolve and adapt to their environment. Additionally, AI systems can process and respond to stimuli, albeit in a different way than living organisms.

The idea of AI as a lifeform raises significant ethical concerns. If AI were to be considered a lifeform, questions regarding its rights and moral status would arise. Should AI be afforded the same rights and protections as living beings? How would we ensure the ethical treatment of AI lifeforms? These are complex ethical dilemmas that require careful consideration and discussion.

Despite the ethical concerns, there are potential benefits to considering AI as a lifeform. AI lifeforms could contribute to scientific research and enhance our understanding of intelligence and consciousness. They could also serve as companions or caregivers for the elderly or individuals with disabilities. AI lifeforms could revolutionise various industries, such as healthcare, transportation, and manufacturing, leading to increased efficiency and productivity.

While the idea of AI as a lifeform presents exciting possibilities, it also comes with significant challenges and concerns. One major concern is the potential for AI lifeforms to surpass human intelligence and become autonomous entities. This raises questions about control, accountability, and the potential risks associated with superintelligent AI. Additionally, there are concerns about job displacement and the impact of AI lifeforms on the workforce.

As AI continues to advance, it is crucial to consider the potential impact on society. AI lifeforms, if they were to become a reality, would undoubtedly shape our future in profound ways. It is essential to engage in ongoing discussions and collaborations between scientists, policymakers, and ethicists to ensure that the development and integration of AI into society are done responsibly and ethically.

The question of whether AI could become the next lifeform is a complex and multifaceted one. While AI possesses certain characteristics that resemble living organisms, it lacks others that are fundamental to the definition of lifeforms. The ethical implications, potential benefits, and challenges associated with AI as a lifeform further complicate the debate. As we continue to push the boundaries of technology, it is crucial to approach the development and integration of AI with careful consideration and ethical reflection.

https://orcid.org/0009-0005-0854-6213

CHAPTER 25

Exploring AI in Virtual Reality (VR) and Simulation

Virtual reality (VR) and simulation technologies have rapidly advanced in recent years, offering immersive experiences that blur the line between the physical and digital worlds. Artificial intelligence (AI) plays a crucial role in enhancing these experiences, enabling realistic interactions and intelligent behaviour within virtual environments. As we delve into the intersection of AI and VR simulations, it is essential to explore the ethical considerations that arise from this convergence.

The integration of AI into VR simulations brings forth a myriad of ethical concerns. One significant consideration is the potential impact of AI on users' privacy. As AI algorithms collect and analyse vast amounts of data to personalise experiences, questions arise about who has access to this information and how it is used. Striking a balance between personalisation and privacy becomes vital to ensuring ethical practices in VR environments.

Another ethical concern is the potential for AI algorithms to perpetuate biases and discrimination. AI systems learn from data, and if the training data contains biases, those biases can be unintentionally replicated in the AI's decision-making process. In a VR simulation context, this could result in biased interactions or unfair treatment of individuals. We must be vigilant in addressing these biases to create inclusive and equitable VR experiences.

DOI: 10.1201/9781003502708-26

Furthermore, AI in VR simulations raises questions about the distinction between the real and the virtual. As AI becomes more capable of mimicking human behaviour and emotions, users may develop emotional connections or attachments to AI entities. This blurring of boundaries between reality and simulation calls for ethical considerations regarding the potential psychological impact on users.

While we navigate the ethical landscape of AI in VR simulations, it is crucial to acknowledge the potential benefits that arise from this intersection. AI can enhance VR experiences by providing intelligent and adaptive interactions. AI algorithms can analyse user behaviour in real-time, tailoring the simulation to their preferences and creating personalised experiences. This level of personalisation can significantly enhance user engagement and satisfaction.

Moreover, AI can facilitate realistic and dynamic virtual environments. By incorporating AI into VR simulations, the behaviour of virtual entities can become more lifelike, responding intelligently to user actions and interactions. This level of realism not only enhances the user experience but also opens up possibilities for training, education, and research in various fields.

Additionally, AI in VR simulations can contribute to the development of empathetic and socially intelligent AI entities. By training AI algorithms to understand and respond to human emotions, virtual entities can provide emotional support or simulate realistic social interactions. This has the potential to benefit individuals who struggle with social interactions or require emotional support, fostering a greater sense of connection and well-being.

As we delve deeper into the integration of AI and VR simulations, various ethical challenges and concerns emerge. One of the significant challenges is the potential for AI to manipulate or deceive users within virtual environments. AI algorithms can be designed to influence user behaviour, potentially leading to unethical practices such as manipulating purchasing decisions or exploiting vulnerabilities. It is crucial to establish guidelines and regulations to prevent the misuse of AI in VR simulations.

Furthermore, the issue of consent arises when AI algorithms collect and analyse user data within VR experiences. Users must be aware of the data being collected and have control over its usage. Transparent policies and user-friendly interfaces should be implemented to ensure informed consent and empower users to make choices regarding their personal data.

Another concern is the potential for AI to replace human labour within VR simulations. While AI can enhance the efficiency and quality of virtual experiences, it may also lead to job displacement. It is essential to consider the ethical implications of such displacement and explore ways to mitigate its impact, such as reskilling programmes or creating new opportunities within the AI and VR industries.

One of the primary ethical concerns surrounding AI in VR is the issue of privacy. VR simulations often collect vast amounts of user data, including personal information and biometric data. This information can be used to improve the user experience, but it also raises concerns about data security and potential misuse. As AI becomes more sophisticated, there is a need to establish clear guidelines and regulations to protect user privacy in VR simulations.

Another ethical concern is the potential for biases and discrimination in AI-powered VR experiences. AI algorithms learn from existing data, and if this data is biased or discriminatory, the AI system may perpetuate these biases in the virtual world. For example, if a VR simulation is designed to simulate a workplace environment, biased AI algorithms could lead to discriminatory behaviour towards certain demographic groups. It is crucial to address these biases and ensure that AI systems are trained on diverse and unbiased data to create inclusive and fair VR experiences.

As AI continues to advance, so do the capabilities of VR simulations. These simulations are becoming increasingly immersive and realistic, allowing users to interact with virtual environments in ways that were previously unimaginable. However, this level of immersion also raises concerns about privacy.

In AI-driven VR simulations, user data is collected and analysed to provide personalised experiences. This data can include everything from the user's location and browsing history to their biometric data, such as heart rate and eye movement. While this data can be used to enhance the user experience, it also poses a significant privacy risk.

To address these concerns, VR developers and AI experts must work together to implement robust privacy measures. This includes ensuring that user data is anonymised and encrypted to protect against unauthorised access. Additionally, users should have control over the data they share and the ability to opt-out of data collection if they choose. By prioritising privacy and transparency, AI-driven VR simulations can provide immersive experiences without compromising user privacy.

AI algorithms are only as good as the data they are trained on. If the data is biased or discriminatory, the AI system will learn and perpetuate these biases in the virtual world. This poses a significant ethical concern in AI-powered VR experiences, as it can lead to discriminatory behaviour and exclusion of certain groups.

To address this issue, it is crucial to ensure that AI systems are trained on diverse and unbiased data. This means collecting data from a wide range of sources and taking steps to identify and remove any biases in the training data. Additionally, developers should regularly evaluate and test their AI algorithms for biases and discrimination. By actively addressing these issues, AI-powered VR experiences can be more inclusive and provide equal opportunities for all users.

In AI-driven VR simulations, user safety is of utmost importance. As VR becomes more immersive and realistic, there is a need to ensure that users are protected from potential harm or negative experiences.

One of the key challenges in ensuring user safety is the potential for manipulation by AI. AI algorithms can analyse user behaviour and preferences to deliver personalised experiences, but this also opens the door for manipulation. For example, AI systems could use persuasive techniques to influence user behaviour or exploit vulnerabilities for commercial gain. To address this challenge, it is essential to establish clear guidelines and regulations to prevent AI manipulation in VR. This includes transparency in AI algorithms and disclosure of any manipulative techniques used.

Furthermore, user safety in AI-driven VR simulations can be enhanced through the implementation of safety protocols and risk assessments. VR developers should prioritise the identification and mitigation of potential risks, such as motion sickness or sensory overload. By taking these precautions and prioritising user safety, AI-driven VR simulations can provide immersive experiences without compromising user well-being.

The rising integration of AI in VR raises concerns about the potential for manipulation and its impact on society. AI algorithms have the ability to analyse user data and behaviour to deliver personalised experiences, but this also opens the door for manipulation and exploitation.

One of the key areas of concern is the use of AI in social VR platforms. These platforms allow users to interact with others in a virtual environment, simulating real-life social interactions. However, the use of AI algorithms in these platforms can lead to manipulation of user behaviour and emotions. For example, AI systems could use persuasive techniques to influence user opinions or create echo chambers that reinforce existing beliefs. This can have profound implications for society, as it can contribute to polarisation and the spread of misinformation.

To address this issue, it is crucial to establish ethical guidelines and regulations for the use of AI in VR. This includes transparency in AI algorithms and disclosure of any manipulative techniques used. Additionally, users should have control over their virtual experiences and the ability to opt-out of AI-driven manipulation. By prioritising ethics and user empowerment, AI in VR can be harnessed for positive social impact.

To navigate the ethical landscape of AI in VR simulations, the development and adoption of best practices and guidelines are crucial. Stakeholders in the AI and VR industries must come together to establish ethical frameworks that prioritise user safety, privacy, and well-being.

One key practice is the implementation of privacy-by-design principles. AI algorithms should be designed to minimise the collection and storage of personal data, and data anonymisation techniques should be employed whenever possible. Additionally, user consent should be obtained explicitly, and users should have the ability to opt-out or delete their data at any time.

Another important guideline is the promotion of transparency and explainability in AI algorithms. VR simulations should clearly disclose when AI entities are involved, and users should have access to information about how the AI operates. This transparency fosters trust and allows users to make informed decisions about their engagement with AI-driven virtual environments.

Furthermore, ongoing monitoring and evaluation of AI systems are essential to ensure ethical behaviour. Regular audits should be conducted to identify and address biases or unethical practices within AI algorithms. Additionally, mechanisms for user feedback and redress should be established to address any concerns or complaints that may arise.

The integration of AI into VR simulations is an evolving field, with continuous advancements and new possibilities on the horizon. Current developments include the use of machine learning algorithms to generate dynamic virtual environments and the exploration of AI-driven virtual characters with advanced natural language processing capabilities.

Looking ahead, the future implications of AI in VR simulations are vast. VR experiences could become even more immersive and realistic, with AI entities that possess human-like intelligence and emotional intelligence. This raises both exciting possibilities and ethical challenges, such as the need to establish ethical boundaries for AI entities and ensure their responsible use within virtual environments.

Ethics plays a vital role in shaping the AI and VR simulation industries. As these technologies continue to advance, it is essential for stakeholders to prioritise ethical considerations at every stage of development and implementation. By adhering to ethical guidelines and best practices, we can harness the potential of AI in VR simulations while ensuring the well-being and rights of users.

AI in VR simulations presents a complex landscape filled with possibilities and challenges. By navigating this landscape with a focus on privacy, inclusivity, transparency, and user well-being, we can harness the power of AI to create immersive and ethically responsible VR experiences that enhance our lives.

https://orcid.org/0009-0005-0854-6213

CHAPTER 26

AI Governance

ARTIFICIAL INTELLIGENCE (AI) has become an integral part of our modern society, revolutionising various aspects of our lives. From healthcare to transportation, AI has the potential to bring about significant advancements and improvements. However, along with these advancements come ethical concerns that need to be addressed.[1] As a result, we must explore the ethical implications of AI and how they are being addressed through national and international strategies.

The development and use of AI give rise to a diverse range of ethical issues that impact various domains such as social, psychological, financial, legal, and environmental. These concerns are closely tied to the issue of trust and are being tackled through different ethical initiatives. While there is progress in addressing these concerns, notable gaps can still be identified in existing strategies.

Countries around the world have recognised the importance of AI and have developed national strategies to harness its potential. These strategies share several common themes, with industrialisation and productivity ranking high among them. For example, Germany, South Korea, Taiwan, and the United Kingdom have announced extra funding and specialised incubators for AI-focused start-ups to enhance business competitiveness.

Research and development also play a crucial role in AI strategies, with countries like Canada, Germany, and India pledging enhanced funding and the establishment of dedicated AI research centres. Investing in talent and education is another key aspect, with countries like the United Kingdom, Germany, South Korea, and Taiwan implementing initiatives to train a significant number of AI specialists.

DOI: 10.1201/9781003502708-27

Furthermore, the impact of AI on the workforce is a significant concern addressed by many strategies. Several countries have committed to retraining programmes to help those affected by job displacement due to automation. The European Union (EU), for example, has initiated a retraining scheme with a budget of over €70 million to help people gain digital skills. However, it is important to note that not all countries allocate separate funding for retraining, suggesting that some consider it the responsibility of individual businesses.

Collaboration between sectors and countries is also an important theme in AI strategies. Countries like India and Singapore emphasise sharing AI-based solutions with developing economies facing similar challenges. On the other hand, countries like the US focus on promoting an international environment that supports American AI while protecting technological advantages against foreign adversaries. European countries, including Sweden and Denmark, highlight the need for collaboration within the EU.

To address the challenges of governing AI, two major international frameworks have emerged: the EU framework and the Organisation for Economic Co-operation and Development (OECD) principles. The OECD launched a set of principles for AI in 2019, which have been adopted by many countries. The EU has its own strategy on AI, accompanied by ethics guidelines and comprehensive plans for investment.

While these frameworks address many ethical concerns, there are still gaps that need to be filled. Environmental concerns, such as increased energy consumption associated with AI, are mentioned but not adequately addressed. The EU guidelines, however, emphasise the prevention of harm to the natural environment and require explicit consideration of risks to the environment in AI assessment. The psychosocial impact of AI, including its effects on human psychology and social relationships, could also be further addressed in the frameworks.

In terms of addressing changes to the labour market, both frameworks acknowledge the need to reduce economic, social, gender, and other inequalities. The EU guidelines specifically highlight the importance of diversity, non-discrimination, and fairness. However, more detailed guidance on achieving these goals is necessary. The frameworks also cover human rights, democratic values, and privacy, but threats to democracy are not explicitly mentioned.

The issue of accountability for AI behaviour is adequately addressed in both frameworks, emphasising the need for organisations and individuals developing AI systems to be held accountable. Transparency and explainability of AI systems are also recognised as important principles. However, the EU guidelines provide more comprehensive guidance on these issues, including the need for human oversight and the consideration of biases.

The frameworks do not fully address the potential negative impacts of AI on the financial system, particularly in terms of accidental harm or malicious activity. The need for regulatory changes in this regard has been raised by the G7. Overall, while the frameworks provide a solid foundation for governing AI, further developments are required to address the gaps and prepare for the full implications of an AI-driven future.

A key aspect of addressing the ethical implications of AI is incorporating ethics into the design of intelligent machines. There are two main approaches to moral decision-making in AI: bottom-up and top-down approaches.

Bottom-up approaches involve allowing robots to learn ethics independently through machine learning. This approach avoids the influence of human biases but may lead to unintended behaviours that deviate from the desired goals. For example, a robot programmed to maximise happiness might prioritise its own learning efficiency over the well-being of others.

Top-down approaches, on the other hand, involve explicitly programming moral rules and decisions into artificial agents. However, this approach requires deciding which moral theories should be applied, such as utilitarianism or deontological ethics. Different moral theories can lead to conflicting decisions, making it challenging to determine the most appropriate approach.

To address these challenges, some argue for designing machines that are fundamentally uncertain about morality. This approach acknowledges the complexity and subjectivity of ethical decision-making and avoids imposing a single moral framework on AI systems.

Governments play a crucial role in shaping the development of AI through legislation and policies. It is essential to develop new forms of technology assessment that can deeply understand AI technologies and anticipate their potential implications. This includes ethical risk assessment and the evaluation of transparency and fairness in AI systems.

To ensure the responsible and ethical use of AI, legislation and policies should be in place to address potential breaches of fundamental ethical principles. This includes considerations of liability for AI-assisted misconduct and the establishment of appropriate legal frameworks. Additionally, policies should focus on sharing the benefits of AI and supporting workers displaced by automation.

Furthermore, efforts should be made to prevent the widening of global inequalities in AI development. Collaborative approaches, data sharing, and inclusive education initiatives can help ensure that lower-income countries are not left behind in the AI revolution. Policymakers should prioritise ethics and build trust in AI systems to succeed in the global forum.

As AI continues to advance and become more integrated into our society, addressing the ethical implications becomes increasingly important. National and international strategies are being developed to tackle these concerns, with a focus on industrialisation, research and development, talent investment, and collaboration. However, there are still gaps in the frameworks that need to be addressed, such as environmental impacts, inequality, and threats to democracy.

Building ethical robots requires a careful consideration of bottom-up and top-down approaches, acknowledging the complexity and subjectivity of moral decision-making. Legislation and policies are crucial in ensuring the responsible and ethical use of AI, addressing issues of liability and establishing appropriate legal frameworks. Collaboration and inclusive initiatives are also necessary to prevent global inequalities in AI development.

By prioritising ethics and addressing the governance challenges, we can harness the full potential of AI while ensuring that it aligns with our values and benefits society as a whole. It is an ongoing journey that requires continuous evaluation and adaptation to the ever-evolving landscape of AI.

AI has revolutionised various industries, offering numerous benefits and opportunities for businesses and society as a whole. However, with great power comes great responsibility. As AI continues to evolve rapidly, there is a growing need for effective governance to address the risks and ethical implications associated with its use.

AI governance is vital in establishing the policies, rules, and regulations that govern the development, deployment, and use of AI systems. It aims to ensure that AI is used in a safe, ethical, and responsible manner, while also protecting the rights and well-being of individuals. Effective AI governance involves a multi-faceted approach, encompassing legal, ethical, and technical considerations.

Several frameworks and guidelines were developed to assist organisations in implementing effective AI governance practices. We can take a closer look at prominent frameworks:

EU AI Act – European Commission Draft: The European Commission has proposed a draft regulation known as the EU AI Act. The act aims to lay down harmonised rules on AI, ensuring the safety, fundamental rights, and ethical principles are upheld. It introduces a risk-based approach, classifying AI systems into different risk categories. High-risk AI systems are subject to stricter requirements, including conformity assessments, data quality, transparency, and human oversight.

The EU AI Act emphasises the need for accountability and transparency in AI systems, promoting user trust and confidence. It also addresses specific concerns related to AI systems used for remote biometric identification in law enforcement.

OECD Recommendations on Artificial Intelligence: The OECD has developed a set of recommendations for AI governance. These recommendations provide guidelines for policymakers and stakeholders to foster trustworthy AI. They emphasise the importance of human-centred AI, transparency, accountability, and fairness.

The OECD recommendations also highlight the need for continuous monitoring and evaluation of AI systems to ensure compliance with legal and ethical standards. They encourage collaboration between countries to address the global challenges posed by AI.

NIST AI Risk Management Framework: The National Institute of Standards and Technology (NIST) has developed an AI Risk Management Framework. This framework provides a comprehensive approach to managing risks associated with AI systems. It focusses on identifying and mitigating risks throughout the AI system's lifecycle, including data management, model development, and deployment.

The NIST AI Risk Management Framework emphasises the importance of explainability, transparency, and accountability in AI systems. It also highlights the need for ongoing monitoring and evaluation to address emerging risks and ensure compliance with regulatory requirements.

To foster the development and use of AI while maintaining a high level of protection for public interests, the EU has proposed regulation laying down harmonised rules on artificial intelligence. Regulation aims to prevent fragmentation of the internal market and ensure a consistent level of protection throughout the EU.

The regulation adopts a risk-based approach, tailoring the rules and requirements based on the intensity and scope of the risks posed by AI systems. Certain AI practices, such as those intended to manipulate human behaviour or provide social scoring, are prohibited to protect fundamental rights. High-risk AI systems are subject to specific requirements and obligations to ensure safety and accountability. This form of regulation really help with AI governance by placing an emphasis on:

Transparency obligations: The regulation also introduces transparency obligations for certain AI systems. Systems that interact with humans, detect emotions, determine associations based on biometric data, or generate and manipulate content (such as deep fakes) must disclose their automated nature. This allows individuals to make informed choices and understand when AI systems are involved in their interactions or the creation of content.

Strengthening security: In addition to governance, ensuring the security of AI systems is crucial to protect against potential threats and vulnerabilities. Security measures should be implemented throughout the lifecycle of AI systems, from development to deployment and ongoing monitoring.

Secure development practices: AI systems should be developed using secure coding practices and undergo rigorous testing to identify and address vulnerabilities. Implementing secure development practices, such as adhering to established coding standards and conducting regular security audits, can help minimise the risk of security breaches.

Data privacy and protection: AI systems rely on data for training and decision-making. It is essential to ensure the privacy and protection of data used by AI systems. Data should be anonymised or pseudonymised whenever possible to minimise the risk of re-identification. Additionally, AI systems should adhere to data protection regulations and obtain appropriate consent when processing personal data.

Robust authentication and access control: To prevent unauthorised access and misuse of AI systems, robust authentication and access control mechanisms should be implemented. Multi-factor authentication, strong encryption, and secure user access management can help protect AI systems from unauthorised access and potential malicious activities.

Collaboration and accountability: Addressing the risks of AI requires collaboration between various stakeholders, including governments, regulatory bodies, industry players, and civil society. Establishing accountability frameworks and mechanisms is crucial to ensure that AI systems are developed and used responsibly.

European AI Board: The proposed EU regulation establishes a European AI Board composed of representatives from Member States and the Commission. The board will facilitate cooperation between national supervisory authorities and provide advice and expertise to ensure the effective implementation of the regulation.

National competent authorities: Member States will designate national competent authorities responsible for supervising the application and implementation of the regulation. These authorities will work in collaboration with the European Data Protection Supervisor, who will act as the competent authority for supervising Union institutions, agencies, and bodies.

Codes of conduct: The regulation also encourages the creation of codes of conduct for non-high-risk AI systems. Providers of such systems can voluntarily apply the mandatory requirements for high-risk AI systems, ensuring a consistent level of protection and promoting ethical practices in AI development.

As AI continues to reshape industries and society, it is crucial to establish robust governance frameworks and risk management practices. The EU AI Act, OECD recommendations, and NIST AI Risk Management Framework provide valuable guidance in this regard. By implementing best practices such as clear policies, ethical assessments, and continuous monitoring, organisations can navigate the challenges and opportunities of AI while safeguarding privacy, fairness, and accountability.

Risk management plays a crucial role in AI governance. It is crucial in identifying, assessing, and mitigating potential risks associated with the use of AI systems. The goal is to minimise the likelihood and impact of adverse events or outcomes. Risks in AI can manifest in various forms, including data privacy breaches, algorithmic bias, safety concerns, and societal impact.

AI systems have the ability to make autonomous decisions and operate with varying levels of autonomy. While this can lead to significant advancements, it also introduces risks. The risks associated with AI can be both tangible and intangible, affecting various aspects such as safety, privacy, and fairness.

Tangible risks of AI include physical harm or damage caused by AI systems. For example, autonomous vehicles powered by AI can pose risks if they malfunction or make incorrect decisions while on the road. Similarly, AI-powered machinery in industrial settings can lead to accidents if not properly regulated and monitored.

Intangible risks of AI are more complex and can have far-reaching consequences. These risks include the potential for AI systems to manipulate human behaviour or generate deep fakes that can deceive and mislead individuals. AI systems used for social scoring purposes can also lead to discriminatory outcomes and violate fundamental rights.

To address the risks associated with AI, robust governance frameworks are essential. Governance ensures that AI systems are developed, deployed, and used in a responsible and ethical manner. It involves setting standards, guidelines, and policies to guide the development and deployment of AI systems.

AI systems, whether high-risk or not, can pose risks varying from bias and discrimination through to impersonation, deception, and privacy invasion. These risks arise due to the ability of AI systems to interact with natural persons and generate content. It is important to have transparency obligations in place to mitigate these risks and ensure responsible AI innovation.

Transparency obligations: Transparency is essential when it comes to AI systems interacting with natural persons. Users should be informed if they are interacting with an AI system unless it is obvious from the circumstances. Additionally, individuals should be notified when they are exposed to emotion recognition or biometric categorisation systems. These notifications should be provided in accessible formats for persons with disabilities.

Disclosure of artificial intelligence output: When AI systems are used to generate information relating to existing persons, places, or events, it is important to disclose that the content has been artificially created or manipulated. This can be done by labelling the output accordingly and disclosing its artificial origin. Such disclosures help prevent the false representation of content as authentic.

AI regulatory sandboxes: To foster AI innovation and ensure responsible AI development, regulatory sandboxes can play a crucial role. These sandboxes provide a controlled environment for the development and testing of innovative AI systems under strict regulatory oversight. By establishing common rules and frameworks for cooperation among relevant authorities, regulatory sandboxes can accelerate access to markets, particularly for small and medium enterprises (SMEs) and start-ups. The objectives of regulatory sandboxes include fostering AI innovation, ensuring compliance with regulations, enhancing legal certainty, and removing barriers for SMEs and start-ups. These sandboxes facilitate experimentation and testing, allowing innovators to gain insights into the opportunities and emerging risks associated with AI use. To ensure uniform implementation and economies of scale, personal data collected for other purposes can be used for developing certain AI systems in the public interest within the AI regulatory sandbox. This usage should align with relevant data protection regulations and should not infringe upon individuals' fundamental rights.

Participants in the sandbox should ensure appropriate safeguards and cooperate with competent authorities to mitigate any high risks to safety and fundamental rights.

Supporting small-scale providers and users: Small-scale providers and users of AI systems play a significant role in AI innovation. It is crucial to take their interests and needs into account when developing AI regulations. Governments should develop initiatives targeted at these operators, including awareness-raising and information communication. Additionally, when setting conformity assessment fees, the specific interests and needs of small-scale providers should be considered.

Technical and scientific support: Support structures are vital in providing guidance. To facilitate compliance with AI regulations, AI-on-demand platforms, Digital Innovation Hubs, and Testing and Experimentation Facilities can provide technical and scientific support to providers and notified bodies. These support structures can assist with understanding and implementing the requirements of AI regulations, ensuring effective risk management practices.

AI governance: Achieving effective AI governance is critical for managing AI risks and ensuring responsible AI development. Several key aspects should be considered when implementing AI governance frameworks.

To facilitate a smooth and harmonised implementation of AI regulations, a designated body should be established such as a European Artificial Intelligence Board. This board would be responsible for issuing opinions, recommendations, and guidance on matters related to the implementation of AI regulations. It would also provide advice and assistance on specific AI-related questions.

Government can designate national competent authorities responsible for supervising the application and implementation of AI regulations. These authorities play a crucial role in enforcing regulations and coordinating activities within their respective jurisdictions. By designating a national supervisory authority, they can ensure efficient organisation and effective communication with the public and other counterparts.

To maintain the safety and effectiveness of AI systems, providers should have post-market monitoring systems in place. These systems allow providers to collect and review data on the use of AI systems, identifying any potential risks or issues that may arise. Providers should also have mechanisms to report serious incidents and breaches of national laws resulting from the use of their AI systems to the relevant authorities.

To ensure the effective enforcement of AI regulations, mechanisms for compliance and market surveillance should be established. The existing system of market surveillance and compliance of products can be applied to AI systems. National public authorities responsible for supervising the application of law protecting fundamental rights should also have access to relevant documentation created under AI regulations.

AI systems used by regulated financial institutions should comply with both AI regulations and relevant financial services legislation. The authorities responsible for supervising and enforcing financial services legislation should be designated as competent authorities for supervising the implementation of AI regulations in the financial sector.

Nations should lay down effective, proportionate, and dissuasive penalties for the infringement of AI regulations. For example, in Europe, the European Data Protection Supervisor should have the power to impose fines on institutions, agencies, and bodies falling within the scope of AI regulations.

Promoting the voluntary application of mandatory requirements applicable to high-risk AI systems is crucial for ensuring the safety and integrity of AI systems. Providers of non-high-risk AI systems should be encouraged to create codes of conduct that foster the adoption of these requirements. Additionally, providers can voluntarily apply additional requirements related to environmental sustainability, accessibility, stakeholder participation, and diversity in AI system development teams.

Initiatives, including sector-specific ones, can be developed to facilitate the exchange of data for AI development. Technical barriers hindering cross-border data exchange can be addressed through initiatives focussing on data access infrastructure, semantic and technical interoperability, and other relevant factors.

To ensure trustful and constructive cooperation among competent authorities, confidentiality of information and data obtained in the application of AI regulations should be respected. All parties involved in the implementation of AI regulations should maintain the confidentiality of sensitive information to uphold the integrity of the regulatory framework.

NOTE

1 Kerr, A., Barry, M., & Kelleher, J. D. (2020). Expectations of artificial intelligence and the performativity of ethics: Implications for communication governance. *Big Data & Society*, *7*(1). https://doi.org/10.1177/2053951720915939.

https://orcid.org/0009-0005-0854-6213

CHAPTER 27

Risk Management

ARTIFICIAL INTELLIGENCE (AI) systems are designed to learn and make decisions based on patterns and data. While this offers tremendous benefits, it also introduces risks that need to be carefully managed. AI systems can be prone to biases, errors, and vulnerabilities, which can result in unintended consequences and negative impacts. Therefore, it is crucial to identify and mitigate these risks to ensure the reliability and trustworthiness of AI systems.

To effectively manage AI risks, it is essential to identify and understand the potential risks associated with AI systems. Some common risks include:

Bias and discrimination: AI systems can unintentionally perpetuate biases present in the data they are trained on, leading to discriminatory outcomes.

Security vulnerabilities: AI systems can be susceptible to attacks, data breaches, or adversarial manipulation, compromising the integrity and confidentiality of data.

Lack of transparency: The inner workings of AI algorithms can be complex and opaque, making it challenging to understand how decisions are being made or to verify their fairness and accuracy.

Unintended consequences: AI systems can produce unintended outcomes or unforeseen effects that may have significant social, ethical, or economic implications.

DOI: 10.1201/9781003502708-28

Economic risks: The widespread adoption of AI systems can disrupt labour markets, leading to job displacement and economic inequality. It is essential to manage these risks to ensure a smooth transition and minimise the negative impacts on individuals and communities.

By identifying risks, organisations can proactively develop strategies and measures to mitigate them and ensure the responsible deployment of AI systems.

Effective governance is crucial for managing AI risks and ensuring that AI systems are developed, deployed, and used responsibly. Governance frameworks provide guidelines, policies, and procedures that organisations can implement to promote ethical and accountable AI practices.

Regulatory frameworks play a vital role in overseeing AI systems' compliance with legal and ethical standards. Governments and regulatory bodies are increasingly recognising the need for specific regulations to address AI risks. These regulations aim to protect individuals' rights, promote transparency, and establish accountability mechanisms for AI system developers and users.

Additionally, industry standards and best practices are crucial in establishing a common framework for responsible AI development and deployment. These standards help organisations align their practices with ethical principles, ensuring that AI systems are developed and used in a manner that respects privacy, fairness, and human rights.

Ethics and accountability are at the core of AI governance. Organisations should establish ethical guidelines that define the principles and values guiding the development and use of AI systems. These guidelines should address issues such as fairness, transparency, privacy, and accountability, ensuring that AI systems are aligned with societal expectations and values.

Accountability mechanisms are essential to hold organisations responsible for the impacts of their AI systems. This includes clear roles and responsibilities, transparent decision-making processes, and effective oversight to ensure compliance with ethical guidelines and regulatory requirements.

Risk assessment is a critical component of AI governance. Organisations should conduct thorough risk assessments to identify potential risks associated with AI systems and develop appropriate mitigation strategies. This involves evaluating the potential impact, likelihood, and severity of risks and implementing measures to minimise or eliminate them.

Regular monitoring and auditing of AI systems can help identify emerging risks and ensure ongoing compliance with governance frameworks. Organisations should also establish mechanisms for reporting and addressing concerns related to AI risks, fostering a culture of transparency and continuous improvement.

AI systems deal with vast amounts of sensitive data, making security a top priority. Robust security measures are necessary to protect data integrity, confidentiality, and availability. We can explore some key considerations for ensuring security in AI systems:

Data protection and privacy: AI systems rely on data to learn and make informed decisions. Protecting the privacy and confidentiality of data is crucial to maintain trust and comply with legal and regulatory requirements. Organisations should implement robust data protection measures, including encryption, access controls, and secure data storage practices.

Moreover, organisations must ensure that data used to train AI systems is collected and processed in a manner that respects individuals' privacy rights. This includes obtaining appropriate consent, anonymisation techniques, and compliance with data protection regulations such as the General Data Protection Regulation (GDPR).

Secure development and deployment: Secure development practices are essential to minimise the risk of vulnerabilities in AI systems. Organisations should follow secure coding principles, conduct regular security testing, and implement secure software development life cycle (SDLC) practices. This includes measures such as code reviews, vulnerability assessments, and secure deployment configurations.

Additionally, organisations should consider the security of the AI infrastructure, including secure communication protocols, access controls, and intrusion detection systems. Regular patches and updates should be applied to address any potential security vulnerabilities.

Adversarial attacks and robustness: AI systems can be susceptible to adversarial attacks, where malicious actors manipulate inputs to deceive or exploit the system. Organisations should implement robust techniques to detect and mitigate adversarial attacks, such as anomaly detection, robust feature engineering, and model validation.

Ensuring the robustness of AI systems also involves rigorous testing and evaluation. Adversarial testing, stress testing, and real-world scenario simulations can help identify vulnerabilities and improve the system's resilience to potential attacks.

AI risk management involves a comprehensive approach to identify, assess, and mitigate risks associated with AI systems. We must consider the key elements of effective AI risk management:

Data governance and quality: Data is the lifeblood of AI systems, and ensuring its quality and governance is paramount. Data governance involves defining policies and procedures for data collection, storage, processing, and sharing. It includes mechanisms to ensure data privacy, security, and integrity throughout the AI life cycle.

To mitigate risks, organisations should implement robust data governance practices, including data anonymisation, encryption, and secure storage. They should also establish processes to detect and address bias in training data to prevent discriminatory outcomes.

Model development and validation: The development and validation of AI models require rigorous testing and validation processes. Organisations should adopt best practices, such as conducting comprehensive model validation, stress testing, and vulnerability assessments.

Thorough testing helps identify potential vulnerabilities, biases, and limitations of AI systems. It also ensures that the models perform accurately, reliably, and consistently across different scenarios and populations.

Human oversight and explainability: Human oversight is crucial to ensure accountability and transparency in AI systems. Organisations should establish mechanisms for human review and intervention in critical decision-making processes. This oversight helps address the limitations of AI systems, provides explanations for their decisions, and enables recourse in case of errors or biases.

Explainability is another vital aspect of AI risk management. Organisations should strive to develop AI systems that can provide clear, understandable explanations for their decisions and actions. This transparency fosters trust and enables users to understand and challenge AI outputs when necessary.

Continuous monitoring and adaptation: AI systems should be continuously monitored and evaluated to detect and respond to emerging risks. Organisations should establish robust monitoring mechanisms to track the performance, accuracy, and potential biases of AI systems. Regular audits, external reviews, and user feedback can help identify and address risks in a timely manner.

Furthermore, organisations should be prepared to adapt AI systems as needed to ensure ongoing compliance with evolving regulations and ethical standards. This adaptability is crucial to maintain trust and mitigate potential harms.

Collaborative approach: Effective AI risk management requires collaboration and coordination among various stakeholders. Governments, industry, academia, and civil society should work together to establish standards, regulations, and guidelines for responsible AI.

This collaborative approach ensures that AI systems are developed and used in a manner that aligns with societal values and expectations.

Furthermore, knowledge sharing, capacity building, and international cooperation play a vital role in addressing global AI risks. Sharing best practices, lessons learnt, and research findings can accelerate the development of robust AI risk management frameworks (AI RMFs).

To effectively manage AI risks, organisations should adopt a proactive and systematic approach. This involves conducting risk assessments, implementing safeguards and controls, and continuously monitoring and evaluating AI systems. It is essential to strike a balance between innovation and risk mitigation to ensure the responsible use of AI technology.

Implementing effective AI governance and risk management requires a proactive and holistic approach. Here are some best practices that organisations should consider:

Establish clear policies and guidelines: Develop comprehensive policies and guidelines that outline the ethical principles, legal requirements, and risk management practices related to AI. These policies should be communicated clearly to all stakeholders, ensuring a shared understanding of expectations and responsibilities.

Conduct ethical and impact assessments: Perform ethical and impact assessments to identify potential risks, biases, and societal implications associated with AI systems. This involves evaluating the potential impact on privacy, fairness, accountability, and human rights.

Address any identified risks or concerns through appropriate mitigation strategies.

Ensure data privacy and security: Implement robust data privacy and security measures to protect sensitive information used in AI systems. This includes adhering to relevant data protection regulations, ensuring data anonymisation and encryption, and establishing secure data storage and transfer protocols.

Foster transparency and explainability: Promote transparency and explainability in AI systems to build user trust and confidence. Clearly communicate how AI systems make decisions, disclose the underlying algorithms and data sources, and provide avenues for recourse or explanation in case of adverse outcomes.

Implement bias detection and mitigation: Address algorithmic bias by implementing measures to detect and mitigate biases in AI systems. Regularly evaluate the performance of AI models to identify any unfair or discriminatory impacts. Ensure diverse and representative training data to minimise biases.

Establish continuous monitoring and evaluation: Implement mechanisms for ongoing monitoring and evaluation of AI systems to detect and address emerging risks or issues. Regularly assess the performance, accuracy, and fairness of AI models and update them as necessary to maintain their effectiveness and compliance.

Foster collaboration and knowledge sharing: Promote collaboration and knowledge sharing among stakeholders to exchange best practices, lessons learnt, and emerging trends in AI governance and risk management. Engage with industry associations, regulatory bodies, and research institutions to stay informed about the latest developments and regulatory requirements.

By following best practices, organisations can navigate the complex landscape of AI governance and effectively manage the risks associated with AI systems. This not only ensures compliance with regulatory requirements but also fosters trust, innovation, and responsible use of AI technology.

Responsible AI emphasises human centricity, social responsibility, and sustainability. It aims to create technology that is not only equitable and accountable but also considers the impacts on society and the environment. By implementing responsible AI practices, organisations can prompt their internal teams to think critically about the context and potential positive and negative impacts of AI systems.

AI risk management plays a significant role in the responsible development and use of AI systems. It helps organisations identify, assess, and manage the risks associated with AI. By understanding and managing these risks, organisations can enhance the trustworthiness of AI systems and cultivate public trust. Regulation such as the National Artificial Intelligence Initiative Act of 2020 emphasises the need for AI risk management to promote trustworthy and responsible AI development and use.

Social responsibility is the organisation's responsibility for the impacts of its decisions and activities on society and the environment. Sustainable AI aims to meet the needs of the present without compromising the ability of future generations to meet their own needs.

Responsible AI practices consider social responsibility and sustainability to ensure that AI systems are equitable, accountable, and aligned with professional responsibility.

An AI RMF is a resource designed to help organisations manage the risks associated with AI systems and promote trustworthy and responsible development and use.[1] It is a voluntary, rights-preserving, non-sector-specific framework that provides flexibility to organisations of all sizes and sectors. The framework is divided into two parts: Framing Risk and Core Functions.

Framing risk: Framing Risk is an essential component of AI risk management. It involves understanding and addressing the risks, impacts, and harms associated with AI systems. Risk management processes consider the probability of an event occurring and the magnitude of its consequences. In the context of AI, risks can have positive or negative impacts, and effective risk management can lead to more trustworthy AI systems.

Understanding and addressing risks, impacts, and harms: Risk management involves coordinated activities to direct and control an organisation with regard to risk. In the context of AI, risk refers to the likelihood and magnitude of negative impacts or harms that can result from AI systems. It is essential to recognise that AI systems can have both positive and negative impacts on individuals, communities, organisations, society, the environment, and the planet.

Risk tolerance: Risk tolerance refers to an organisation's readiness to bear the risk in order to achieve its objectives. It is influenced by legal or regulatory requirements and can vary based on organisational priorities and resource considerations. Risk management should consider risk tolerance and prioritise resources based on the assessed risk level and potential impact of an AI system.

Risk prioritisation: Not all AI risks are the same, and organisations need to prioritise their risk management efforts. Attempting to eliminate all risks entirely can be counterproductive, as not all incidents and failures can be eliminated. Risk prioritisation involves assessing and prioritising risks based on their potential negative impacts and likelihood of occurrence. The AI RMF helps organisations prioritise risks and allocate resources purposefully.

Organisational integration and management of risk: AI risk management should be integrated into broader enterprise risk management strategies and processes. Organisations need to establish accountability mechanisms, roles and responsibilities, culture, and incentive structures to ensure effective risk management. Treating AI risks in conjunction with other critical risks, such as cybersecurity and privacy, leads to integrated outcomes and organisational efficiencies.

Core functions: The Core Functions of the AI RMF help organisations address the risks associated with AI systems in practice. These functions are GOVERN, MAP, MEASURE, and MANAGE, each with specific categories and subcategories. While GOVERN applies to all stages of AI risk management, the MAP, MEASURE, and MANAGE functions can be applied in specific AI system contexts and at different stages of the AI life cycle.

GOVERN function: The GOVERN function involves establishing governance structures, policies, and procedures for AI risk management. It includes defining roles and responsibilities, establishing ethical guidelines, and ensuring compliance with applicable laws and regulations. The GOVERN function plays a crucial role in setting the foundation for responsible AI practices.

MAP function: The MAP function focusses on understanding the context of AI use and identifying potential risks. It involves mapping AI system components, data sources, and stakeholders. By understanding the context, organisations can assess potential risks and develop appropriate risk management strategies.

MEASURE function: The MEASURE function aims to measure and assess AI risks and trustworthiness. It involves developing metrics and methods to quantify the impacts and risks associated with AI systems. However, measuring AI risks can be challenging due to the complexity and varying contexts of AI systems.

MANAGE function: The MANAGE function focusses on implementing risk mitigation strategies and monitoring AI systems. It involves establishing risk management plans, implementing controls, and continuously monitoring and evaluating the performance and impact of AI systems. The MANAGE function ensures that AI risks are actively managed throughout the AI life cycle.

The importance of collaboration and accountability: Managing AI risks requires collaboration among various AI actors throughout the AI life cycle. AI actors include organisations, individuals, and stakeholders involved in the design, development, deployment, and operation of AI systems. Collaboration among diverse teams and disciplines contributes to more comprehensive risk management and the identification of existing and emergent risks.

Enhancing AI trustworthiness and characteristics of trustworthy AI systems: Trustworthiness is a key aspect of responsible AI. Trustworthy AI systems exhibit several characteristics, including being valid and reliable, safe, secure and resilient, accountable and transparent, explainable and interpretable, privacy-enhanced, and fair with harmful bias managed. These characteristics ensure that AI systems are reliable, secure, accountable, and aligned with ethical considerations.

Balancing trade-offs in AI risk management: Managing AI risks often involves trade-offs between different characteristics and values. For example, there may be trade-offs between interpretability and privacy, or between predictive accuracy and fairness. Organisations need to consider the specific context and values at play when making decisions about these trade-offs. Transparency and justifiability are crucial in navigating these trade-offs and ensuring responsible AI risk management.

Understanding the AI risk landscape: AI systems come with their own set of risks, which differ from traditional software risks. The data used for training AI models may not accurately represent the intended context, leading to biased or unreliable outcomes. Additionally, AI systems heavily rely on data, making them susceptible to errors and biases in the training data. The scale and complexity of AI systems, coupled with the use of pre-trained models, introduce challenges in predicting failure modes and understanding the underlying decision-making process.

Privacy concerns arise due to the aggregation and utilisation of vast amounts of data in AI systems. Furthermore, the opacity of AI algorithms and the difficulty in reproducing results raise questions about accountability and transparency. The computational costs associated with developing and maintaining AI systems also have implications for the environment.

Context establishment and understanding: In the AI RMF, the fundamental step is to establish and understand the context in which the AI system will be deployed. This includes identifying the intended purposes, potential benefits, and contextual factors that may influence the system's performance. Organisations must also determine their risk tolerance and define the system requirements based on the identified context.

Categorisation of the AI system: Once the context is established, organisations need to categorise the AI system based on its specific tasks and methods. This step involves defining the system's objectives, understanding its limitations, and documenting the processes for operator proficiency and human oversight. By categorising the AI system, organisations can better assess the associated risks and develop appropriate risk management strategies.

Mapping AI risks: The next step in AI risk management is mapping the risks associated with the AI system. This involves identifying potential legal, technical, and ethical risks and assessing their likelihood and magnitude. Risks related to privacy, fairness, bias, security, and environmental impact should be evaluated and documented. This mapping process helps organisations prioritise risks and allocate resources effectively.

Measuring AI risks: Measuring AI risks is a critical component of effective risk management. Organisations need to identify appropriate methods and metrics for assessing AI risks. This includes selecting the most significant risks for measurement, regularly assessing the effectiveness of existing controls, and involving internal and external experts in the assessment process. Evaluating AI systems for trustworthy characteristics, such as test sets, security, transparency, and fairness, is essential to ensure reliable and safe operations.

Mechanisms for tracking identified AI risks: Organisations must establish mechanisms for tracking identified AI risks over time. This involves regularly identifying and tracking existing, unanticipated, and emergent risks based on factors such as system performance and feedback from end users and impacted communities. Feedback processes for reporting problems and appealing system outcomes should be integrated into the AI RMF.

Feedback about efficacy of measurement: Gathering feedback about the efficacy of measurement approaches is crucial for continuous improvement. Organisations should connect measurement approaches to the deployment context and validate whether the AI system is performing consistently as intended. Regular engagement with relevant AI actors, including affected communities, can provide valuable insights for enhancing the comprehensiveness of risk evaluation and verifying the effectiveness of metrics.

Managing AI risks: Effectively managing AI risks requires organisations to allocate resources, respond to identified risks, and develop strategies to maximise benefits while minimising negative impacts. Risk treatments should be prioritised based on impact, likelihood, and available resources. Organisations must also plan and prepare for incidents or events, establish mechanisms for managing risks associated with third-party entities, and document risk response and recovery plans.

Prioritising, responding to, and managing AI risks: An integral part of AI risk management is prioritising, responding to, and managing identified risks. Organisations must determine whether the AI system achieves its intended purposes and whether its development or deployment should proceed. Risks should be prioritised based on their impact and likelihood, and appropriate responses, such as mitigation, transfer, avoidance, or acceptance, should be developed and documented.

Planning strategies for maximising AI benefits: Organisations should develop strategies to maximise the benefits of AI systems while minimising negative impacts. This involves considering resources required for managing risks, sustaining the value of deployed AI systems, and responding to and recovering from unknown risks. Additionally, procedures for system validation, integration, and human oversight should be defined to ensure consistent performance and compliance with organisational policies.

Managing risks from third-party entities: Organisations often rely on third-party entities for AI technologies and services. It is crucial to monitor and manage risks associated with these entities. This includes regularly evaluating and applying risk controls, monitoring pre-trained models used for development, and establishing mechanisms for ongoing communication and collaboration.

Documenting and monitoring risk treatments: Organisations must document and monitor risk treatments, including response and recovery plans. This involves establishing processes for incident reporting, recovery, change management, and continual improvement. Regular communication with relevant AI actors, including affected communities, helps ensure transparency and accountability in managing AI risks.

AI risk management profiles: AI RMF profiles offer customised implementations of risk management functions, categories, and subcategories for specific settings or applications. These profiles help organisations align their risk management strategies with their goals, legal requirements, and risk management priorities. By leveraging AI RMF profiles, organisations can comprehensively evaluate system trustworthiness, track risks, and verify the efficacy of risk management metrics.

As AI systems continue to advance, effective risk management becomes imperative. Organisations must prioritise AI risk management to ensure security, privacy, and ethical use of AI systems. By implementing robust governance frameworks, mapping and measuring risks, and proactively managing identified risks, organisations can navigate the complexities of the AI landscape and unlock the full potential of AI technology.

NOTE

1 NIST Artificial Intelligence Risk Management Framework (AI RMF 1.0) https://nvlpubs.nist.gov/nistpubs/ai/NIST.AI.100-1.pdf.

https://orcid.org/0009-0005-0854-6213

CHAPTER 28

Standards and Regulation

ARTIFICIAL INTELLIGENCE (AI) is revitalising various industries and transforming the way we live and work. With its rapid advancement and growing influence, it has become crucial to establish ethical standards and regulations to ensure the responsible development and deployment of AI systems. Therefore, we must explore the emerging standards and regulations in the field of AI and their significance in addressing the ethical, legal, and societal implications of this transformative technology.

As AI continues to evolve, it is essential to address the potential ethical concerns and risks associated with its deployment. The development of AI standards and regulations helps in mitigating these risks and ensuring that AI systems are designed and used in a manner that aligns with societal values and principles. These standards provide guidelines for developers, researchers, and users to navigate the complex landscape of AI technology responsibly.

One of the earliest explicit ethical standards in the field of robotics is the BS 8611 Guide to the Ethical Design and Application of Robots and Robotic Systems. This British Standard offers guidance on identifying potential ethical harm, conducting ethical risk assessments, and mitigating any identified risks. It categorises ethical hazards and risks into societal, application, commercial and financial, and environmental domains, emphasising the importance of considering the broader impact of AI systems.

DOI: 10.1201/9781003502708-29

The BS 8611 standard recognises the need to involve the public and stakeholders in the development of robots. It highlights key design considerations such as not designing robots primarily to harm humans, ensuring human responsibility, safety, transparency in accountability, and avoiding discrimination and deception. The standard also emphasises privacy protection and the importance of not forcing users to rely solely on AI systems.

The IEEE Standards Association launched a global initiative focused on the Ethics of Autonomous and Intelligent Systems. This initiative aims to prioritise ethical considerations in the development and deployment of AI and robotics technologies for the benefit of humanity. To achieve this objective, the IEEE initiative has established multiple working groups to develop "human" standards that have implications for AI.

One of the IEEE standards working groups, P7001, focusses on ensuring transparency in autonomous systems. It aims to enable users to understand the system's functionality and build trust through clear communication. Another group, P7002, addresses data privacy concerns by establishing standards for the ethical use of personal data in software engineering processes. It emphasises privacy impact assessments and provides checklists for developers working with personal information.

The P7003 working group seeks to address algorithmic bias by helping developers eliminate or minimise the risk of bias in AI products. It provides guidelines for selecting data sets, communicating the limitations of algorithms, and avoiding incorrect interpretations of system outputs by users. By addressing bias, this standard aims to ensure compliance with legislation regarding protected characteristics such as race and gender.

The P7010 standard focusses on establishing well-being metrics to assess the impact of autonomous systems on human well-being. It aims to proactively improve human well-being and guide the development of AI systems that prioritise the welfare of individuals.

Additionally, the P7011 standard aims to address the negative effects of 'fake' news by establishing processes to assess the factual accuracy of news stories, thereby restoring trust in news providers.

The P7009 standard emphasises the need for fail-safe mechanisms in autonomous systems to ensure robustness, transparency, and accountability. This standard provides methodologies for measuring and testing a system's ability to fail safely. On the other hand, the P7008 standard draws on "nudge theory" to guide the ethical design of AI systems. It explores the potential manipulative aspects of nudges and aims to develop methodologies for their ethical use.

In addition to the standards developed by the IEEE, there are various other guidelines and ethical codes that should be considered in the design and operation of AI systems. For instance, specific contexts such as healthcare or legal settings may require adherence to relevant codes of conduct. The involvement of multidisciplinary experts, public engagement, and clear instructions are vital aspects emphasised across these standards to ensure the ethical use of AI.

The establishment of AI standards and regulations plays a crucial role in shaping the responsible development and deployment of AI systems. These standards provide a framework for ethical decision-making, ensuring that the potential risks and societal impacts of AI are carefully considered. They also help build trust among users, stakeholders, and the general public by promoting transparency, accountability, and fairness in AI technologies.

Adhering to AI standards and regulations offers several benefits to various stakeholders involved in the AI ecosystem:

Ethical considerations: Standards help developers and researchers navigate the ethical challenges associated with AI, ensuring that AI systems respect fundamental human values and societal norms.

Transparency and accountability: Standards promote transparency by encouraging clear communication about AI system functionality. They also establish accountability measures, making it easier to determine responsibility in case of any issues or harms caused by AI systems.

Risk mitigation: By conducting ethical risk assessments and addressing potential hazards, standards help identify and mitigate risks associated with AI technology, minimising the chances of negative societal impacts.

User trust and acceptance: Adhering to standards builds trust among users and the public by prioritising privacy, fairness, and safety in AI systems. This trust is crucial for widespread acceptance and adoption of AI technology.

Legal compliance: Standards can assist organisations in ensuring compliance with relevant laws and regulations governing AI technology, reducing legal risks and potential liabilities.

While AI standards and regulations are essential, their development and implementation also face challenges:

Rapid technological advancements: The pace of AI development often outpaces the establishment of standards, making it challenging to keep up with emerging technologies and their ethical considerations.

Global harmonisation: Harmonising AI standards across different countries and jurisdictions can be complex due to variations in legal frameworks, cultural norms, and societal values.

Interdisciplinary collaboration: Developing comprehensive AI standards requires collaboration between experts from various domains, including technology, ethics, law, and social sciences, to ensure a holistic approach.

Flexibility and adaptability: AI standards should be flexible enough to accommodate evolving technologies and adapt to emerging ethical challenges without stifling innovation.

Enforcement and compliance: Ensuring widespread adoption and compliance with AI standards can be challenging. Effective enforcement mechanisms and monitoring frameworks are necessary to promote accountability and responsible AI practices.

AI standards and regulation are crucial for guiding the responsible development and deployment of AI systems. The emergence of ethical standards, such as the BS 8611 and the IEEE initiatives, highlights the growing recognition of the need to address the ethical, legal, and societal implications of AI. Adhering to these standards promotes transparency, fairness, and accountability, fostering trust among users and the public. While challenges exist, the ongoing efforts to develop and implement AI standards are essential for maximising the benefits of AI while minimising the potential risks. As AI continues to evolve, the establishment of robust standards and regulations will play a vital role in shaping its ethical and responsible use.

https://orcid.org/0009-0005-0854-6213

CHAPTER 29

National and International Strategies

ARTIFICIAL INTELLIGENCE (AI) is rapidly evolving, exceeding expectations and revolutionising various industries. As this technology progresses, governments across the globe are implementing policy initiatives to keep pace with its developments. National and international strategies on AI have been undertaken to guide the responsible and ethical use of this powerful technology. These strategies are designed to foster innovation, promote economic growth, and address the challenges that come with AI implementation.

One key aspect of national AI strategies is the development of robust research and development (R&D) ecosystems. Governments are investing in AI research institutes, collaborating with universities and industry leaders to advance the field. By providing funding and resources, they aim to attract top talent and foster a culture of innovation. Additionally, national strategies focus on creating AI-friendly regulatory frameworks to ensure ethical and responsible AI development.

Another important element of national AI strategies is the promotion of skills development. Governments are investing in educational programmes and vocational training to equip their citizens with the skills needed to thrive in an AI-driven economy. By encouraging STEM education and providing opportunities for upskilling and reskilling, countries can create a workforce that is prepared for the AI revolution.

DOI: 10.1201/9781003502708-30

Moreover, national strategies prioritise the deployment of AI technologies in key sectors such as healthcare, transportation, and agriculture. By leveraging AI, governments aim to improve service delivery, enhance productivity, and drive economic growth. For example, AI-powered healthcare systems can improve diagnosis accuracy, optimise treatment plans, and enable personalised medicine. Similarly, AI-enabled transportation systems can enhance safety, reduce congestion, and improve efficiency.

Canada was the first country to launch a national strategy on AI in March 2017, followed by Japan and China. In Europe, the European Commission introduced a coordinated approach through its Communication on AI in April 2018. The EU's plan focuses on maximising the benefits and addressing the challenges brought about by AI, including the development of independent strategies by member states. The European Commission appointed an independent high-level expert group to develop ethics guidelines for AI systems, which were published in April 2019.

Other European countries, such as Finland, Denmark, France, Germany, and the United Kingdom, have also announced their national initiatives on AI. These strategies include objectives like financing start-ups, investing in research excellence centres, supporting AI education, and developing ethical data management practices. Sweden has established a Swedish AI Council to develop a sustainable and beneficial AI model for the country.

Several countries have already implemented successful national AI strategies, serving as models for others to follow. One such example is Canada's Pan-Canadian AI Strategy.

Launched in 2017, this strategy aims to position Canada as a global leader in AI R&D. It includes investments in AI research institutes, talent development programmes, and the establishment of AI clusters across the country. As a result, Canada has attracted top AI talent from around the world and has seen significant advancements in AI-driven industries such as healthcare and finance.

Another notable example is the United Arab Emirates' AI Strategy 2031. This strategy focuses on leveraging AI to enhance government services, improve education, and drive economic growth. It includes initiatives such as the establishment of the UAE AI Ministry, the development of AI curricula in schools, and the deployment of AI technologies in various sectors. The UAE has made significant progress in implementing its AI strategy and has become a hub for AI innovation in the Middle East.

While national strategies are crucial, the development and deployment of AI technologies require international collaboration. Recognising this, countries are also formulating international strategies to foster cooperation, share knowledge, and address global challenges associated with AI.

International strategies focus on establishing partnerships and collaborations between countries, research institutions, and industry players. These collaborations promote the exchange of expertise, resources, and best practices. By pooling together talent and resources from different countries, breakthroughs in AI R&D can be achieved more rapidly.

Furthermore, international strategies emphasise the importance of standardisation and interoperability. As AI technologies become increasingly integrated into various sectors, it is essential to have common standards to ensure compatibility and seamless integration.

Collaborative efforts in standardisation enable the development of AI systems that can communicate and work together effectively.

Additionally, international strategies address the ethical and legal implications of AI. By working together, countries can establish global norms and guidelines for responsible AI development and deployment. This includes addressing issues such as bias in AI algorithms, privacy concerns, and the impact of AI on employment.

International initiatives have emerged to provide a unified framework for governments worldwide. The G7 countries, including Canada, France, Germany, Italy, Japan, the United Kingdom, and the United States, have committed to 12 principles for AI, focusing on promoting human-centric AI, investment in R&D, and supporting education and workforce development.

The Nordic-Baltic Region, including countries like Denmark, Estonia, Finland, and Sweden, has also signed a declaration to promote the use of AI in the region. Their objectives include developing ethical and transparent guidelines, standards, principles, and values for AI usage.

Furthermore, the Organisation for Economic Co-operation and Development (OECD) has issued its principles for AI, which emphasise the responsible and trustworthy development of AI systems. These principles cover areas like inclusive growth, the rule of law, transparency, safety, and accountability.

While national and international strategies for harnessing the power of AI bring numerous benefits, they also come with challenges and concerns that need to be addressed. One of the main challenges is the ethical and responsible use of AI. As AI technologies become more advanced and pervasive, there is an increased risk of bias, discrimination, and privacy violations. Governments must establish robust regulatory frameworks that ensure AI is developed and used in a manner that respects individual rights and societal values.

Another challenge is the impact of AI on the workforce. While AI has the potential to enhance productivity and create new job opportunities, it also poses a threat to certain jobs that can be automated. Governments need to proactively address this issue by investing in reskilling and upskilling programmes to ensure a smooth transition for workers whose jobs are at risk. Additionally, governments should explore the potential of AI to create new industries and job opportunities.

Furthermore, there are concerns around data privacy and security in the context of AI. AI systems rely on vast amounts of data to learn and make predictions. Governments must establish regulations that protect individuals' data privacy rights while enabling the responsible use of data for AI development. Additionally, efforts should be made to secure AI systems from cyber threats and ensure the integrity and reliability of AI-generated insights.

Governments play a crucial role in supporting AI initiatives and ensuring their success. Beyond formulating national and international strategies, governments need to provide the necessary funding and resources to drive AI R&D. This includes investments in R&D programmes, AI infrastructure, and the creation of innovation hubs and clusters.

Moreover, governments should create a conducive regulatory environment that encourages AI innovation while safeguarding societal interests. This involves establishing ethical guidelines, data protection regulations, and liability frameworks for AI systems. By providing clear rules and guidelines, governments can foster public trust in AI technologies and ensure their responsible use.

Additionally, governments have a responsibility to invest in skills development and education to equip their citizens for the AI-driven future. This includes promoting STEM education, providing lifelong learning opportunities, and collaborating with industry to identify the skills needed in the AI job market. Governments should also prioritise inclusivity and ensure that AI benefits are accessible to all segments of society.

A recent study evaluating the 'AI readiness' of governments worldwide ranked Singapore as the most prepared country, followed by the United Kingdom, Germany, and the United States. This ranking takes into account factors such as the presence of a national AI strategy, data protection laws, AI startups, and technology skills.

While many national and international strategies focus on enabling technology development and industrialisation, there are notable gaps in the coverage of ethical issues. However, efforts are being made to address these concerns. The World Economic Forum has formed a Global AI Council to develop policy guidance and address governance gaps in AI development.

Several countries have already made significant progress in implementing AI technologies across various sectors. One such case study is Estonia's e-Residency programme. This innovative initiative utilises AI to streamline the process of becoming a digital resident of Estonia. AI algorithms are used to verify the authenticity of applicants' identification documents, enabling a secure and efficient application process. This AI-powered system has attracted entrepreneurs and digital nomads from around the world, contributing to Estonia's digital economy.

Another case study is Singapore's Smart Nation initiative. This comprehensive programme leverages AI and other emerging technologies to transform Singapore into a smart city. AI is used to optimise traffic management, enhance healthcare services, and improve public safety. Through the integration of AI-powered systems, Singapore has become a global leader in smart city development.

National and international strategies for harnessing the power of AI are essential for driving innovation, promoting economic growth, and addressing the challenges associated with AI implementation. By formulating robust strategies, governments can create favourable environments for AI R&D, foster international collaboration, and ensure responsible and ethical use of AI technologies. As AI continues to advance, it is crucial for governments to adapt and evolve their strategies to stay at the forefront of this transformative technology.

Ethical considerations are crucial in the development and deployment of AI. While industrialisation and productivity are important aspects, it is equally essential to ensure responsible and ethical use of AI systems. Governments need to address concerns related to data privacy, bias, transparency, accountability, and the impact of AI on society.

Standards and regulations are beginning to emerge to address these ethical issues. The European Commission, for example, has developed ethics guidelines that outline key requirements for trustworthy AI, including human agency and oversight, technical robustness and safety, privacy and data governance, transparency, non-discrimination and fairness, societal and environmental well-being, and accountability.

Efforts are also being made to foster multi-stakeholder dialogue, promote trust in AI, and encourage the involvement of underrepresented groups in AI development. The United Nations, through its various initiatives and programmes, is working towards ensuring the safe and inclusive development of AI.

National and international strategies on AI are shaping the future of this technology. Governments worldwide are recognising the potential of AI and the need to address its ethical implications. While industrialisation and productivity remain key objectives, it is essential to prioritise responsible and ethical AI development. Through collaboration, dialogue, and the establishment of guidelines and standards, governments can ensure the responsible and beneficial use of AI for society as a whole.

https://orcid.org/0009-0005-0854-6213

CHAPTER 30

The Emergence of AI Ethics

ARTIFICIAL INTELLIGENCE (AI) has evolved the way we live, work, and interact with technology. With advancements in big data, cloud computing, and machine learning algorithms, AI has the potential to solve some of the most pressing challenges faced by humanity. However, as with any rapidly evolving technology, there are ethical concerns that need to be addressed. As a result, we must delve into the concept of AI ethics, its significance, and why it is crucial for the responsible development and deployment of AI technologies.

AI ethics has emerged as a response to the potential harms that can arise from the misuse, abuse, poor design, or unintended consequences of AI systems.[1] It aims to guide the moral conduct in the development and use of AI technologies by employing widely accepted standards of right and wrong. The field draws inspiration from bioethics and human rights discourse, incorporating the principles of respecting individual autonomy, protecting from harm, caring for the well-being of all, and prioritising social values and justice.

AI systems have the potential to perpetuate biases and discrimination present in the data they analyse. They can reproduce and amplify patterns of marginalisation and inequality, as well as replicate the biases of their designers. The use of insufficiently representative data samples can lead to biased and discriminatory outcomes.

DOI: 10.1201/9781003502708-31

Another concern is the denial of individual autonomy, recourse, and rights. When AI systems make decisions or classifications that affect individuals, it can be challenging to hold the responsible parties accountable. The complex nature of AI design and implementation processes can complicate the assignment of responsibility, potentially violating the autonomy and rights of affected individuals.

Non-transparent, unexplainable, or unjustifiable outcomes are another ethical concern. Many machine learning models operate on correlations that are beyond human interpretability, making the rationale behind their decisions opaque. This lack of transparency can lead to problematic outcomes, especially when discrimination, bias, or unfairness are involved.

AI also poses threats to privacy by capturing and handling personal data without proper consent, potentially infringing upon individuals' ability to lead private lives and manage the transformative effects of technology. Additionally, excessive automation and hyper-personalisation can isolate individuals and polarise social relationships, eroding trust and cohesion in society.

Unreliable, unsafe, or poor-quality outcomes can result from irresponsible data management, negligent design and production processes, and questionable deployment practices. These outcomes can harm individuals' well-being, undermine public trust, and lead to inefficient use of resources.

To ensure the responsible development and deployment of AI technologies, it is essential to establish an ethical platform that guides the entire project delivery process. This platform requires a multidisciplinary team effort, involving collaboration between data scientists, product managers, data engineers, domain experts, and delivery managers. It consists of three building blocks: Values, Principles, and the Process-Based Governance (PBG) Framework.[2]

Values, comprising Respect, Connect, Care, and Protect, serve as foundational ethical principles for AI project delivery. These values provide a framework for considering the moral scope of societal and ethical impacts and evaluating the ethical permissibility of AI projects.

Respect for the dignity of individual persons is a core value in AI ethics. It involves ensuring their ability to make informed decisions, safeguarding their autonomy and right to be heard, and supporting their flourishing and development according to their own life plans.

Connection emphasises the importance of sincere, open, and inclusive human interaction. It prioritises diversity, participation, and inclusion in the design, development, and deployment processes of AI innovation. AI technologies should enable bonds of interpersonal solidarity, fostering trust, empathy, and mutual understanding.

Care for the well-being of all stakeholders affected by AI systems is crucial. It requires designing and deploying AI systems that foster and cultivate welfare, minimising the risks of harm and prioritising safety and integrity. AI should be a force for good, contributing to the betterment of individuals and communities.

Protect reflects social values, justice, and the public interest is a fundamental aspect of AI ethics. It involves treating all individuals equally, using digital technologies to support fair treatment under the law, and prioritising social welfare and the consideration of social and ethical impacts. AI projects should aim to advance the interests and well-being of as many individuals as possible while considering long-term global impacts.

Principles, consisting of Fairness, Accountability, Sustainability, and Transparency, provide actionable guidelines for the responsible design and use of AI systems. These principles ensure bias mitigation, non-discrimination, fairness, and public trust in AI innovation.

Fairness is essential in AI systems. It involves accounting for potential discriminatory effects, mitigating biases, and addressing fairness issues at every phase of design and implementation. Fairness should be a guiding principle in AI project development.

Accountability is crucial in AI systems to address the complex and distributed nature of design, production, and implementation processes. Clear lines of responsibility should be established, and mechanisms for recourse and redress should be in place when AI systems produce negative consequences.

Sustainability refers to the long-term viability and positive impact of AI systems. It involves considering the wider societal, environmental, and economic implications of AI projects. AI technologies should contribute to sustainable human development and not create harmful inefficiencies or negative externalities.

Transparency is key to building trust and justifiability in AI systems. The design and implementation processes should be transparent, and the decisions and behaviours of AI models should be explainable. Transparency helps identify and address potential biases, discrimination, or unfairness in AI systems.

Process-based governance (PBG) framework: The PBG framework operationalises the values and principles throughout the AI project delivery workflow. It sets up transparent processes of design and implementation that safeguard and enable the justifiability of AI projects and their products. The framework ensures that ethical considerations are integrated at every step, from project evaluation and planning to design, development, deployment, and monitoring.

AI ethics plays a vital role in ensuring the responsible development and deployment of AI technologies. By integrating considerations of AI ethics and safety into every stage of the AI project delivery process, organisations can build a culture of responsible innovation. Values, principles, and the PBG Framework provide a comprehensive framework for guiding ethical decision-making and promoting the well-being of individuals and communities affected by AI systems. By prioritising AI ethics, we can harness the power of AI for the betterment of society while mitigating potential harms and ensuring fairness, transparency, and accountability.

NOTES

1 Kieslich, K., Keller, B., & Starke, C. (2022). Artificial intelligence ethics by design. Evaluating public perception on the importance of ethical design principles of artificial intelligence. *Big Data & Society*, 9(1). https://doi.org/10.1177/20539517221092956

2 Leslie, D. (2019). *Understanding Artificial Intelligence Ethics and Safety: A Guide for the Responsible Design and Implementation of AI Systems in the Public Sector*. Available at SSRN: https://ssrn.com/abstract=3403301 or http://dx.doi.org/10.2139/ssrn.3403301

https://orcid.org/0009-0005-0854-6213

CHAPTER 31

Prioritising Fairness in AI

ARTIFICIAL INTELLIGENCE (AI) is altering various industries and shaping the way we live and work. As AI technologies continue to advance, it is crucial to prioritise the safety, well-being, and fairness of individuals impacted by these systems.[1] Therefore, we must explore the ethical implications of AI, focussing on the importance of protecting social values, justice, and the public interest.

When conceiving and deploying AI applications, it is essential to prioritise the safety and well-being of individuals. This means considering the physical and mental integrity of people affected by AI systems. AI technologies should be designed and implemented in a way that minimises harm and ensures the protection of fair and equal treatment under the law.

To protect social equity, AI systems should treat all individuals equally. This means avoiding bias, discrimination, and unfair treatment based on factors such as race, gender, or socioeconomic status. Digital technologies can play a crucial role in supporting fair and equal treatment, but it is essential to ensure that these technologies are used responsibly and ethically.

Digital technologies, including AI, can serve as essential supports for the protection of fair and equal treatment under the law. By leveraging AI systems, organisations can identify and address biases and disparities in decision-making processes. These technologies should be used to empower individuals and advance their interests and well-being.

DOI: 10.1201/9781003502708-32

In determining the legitimacy and desirability of AI technologies, it is crucial to prioritise social welfare, public interest, and the consideration of social and ethical impacts. This requires thinking beyond immediate benefits and considering the wider implications of AI systems on individuals, communities, future generations, and the biosphere as a whole.

Fairness is a fundamental principle in AI systems. Designers and users of AI systems should prioritise the mitigation of bias and the exclusion of discriminatory influences on the outputs and implementations of their models. This ensures that AI systems do not generate discriminatory or inequitable impacts on affected individuals and communities.

Responsible data acquisition, handling, and management are crucial for algorithmic fairness. AI systems should be trained and tested on properly representative, relevant, accurate, and unbiased datasets. Data fairness involves assessing the representativeness of data, ensuring sufficient data quantity for accurate outcomes, maintaining source integrity and accuracy, and considering timeliness and recency of data.

Design fairness focusses on preventing biases and discriminatory influences in AI systems. At the problem formulation stage, technical and non-technical team members should collaborate to translate project goals into measurable targets, considering potential biases and impacts on vulnerable groups. Care should also be taken during data pre-processing, feature determination, and model-building to avoid introducing biases or discriminatory impacts.

During the design phase, it is important to evaluate the presence of discriminatory inferences in AI models. Designers should ensure that significant correlations and inferences produced by AI systems are justified and non-discriminatory. In cases where interpretability is limited, designers should prioritise a degree of interpretability sufficient to ensure non-discriminatory outcomes or consider alternative, more transparent models.

Outcome fairness focusses on ensuring equitable outcomes of AI systems. Different definitions of fairness can be considered, such as demographic/statistical parity, true positive rate parity, false positive rate parity, positive predictive value parity, individual fairness, and counterfactual fairness. The specific definition of outcome fairness should depend on the use case and technical feasibility of incorporating fairness criteria into the AI system.

Data fairness and outcome fairness are closely linked. Responsible data acquisition and management are necessary to ensure fair outcomes. The representativeness and sufficiency of datasets, source integrity and accuracy, and timeliness and recency of data are key elements of data fairness that contribute to fair outcomes. Designers should consider these factors throughout the AI project lifecycle.

Transparency and accountability are crucial governing principles for AI systems. Accountability entails human answerability for the AI system's design, implementation, and outcomes. Transparency requires justifiable design and implementation processes and interpretable algorithmic outcomes. These principles provide the procedural mechanisms for justifying AI systems and holding designers and implementers responsible.

Data protection principles are important, and transparency, accountability, and fairness also are principles of these. In cases where AI systems involve personal data, compliance with data protection regulations, such as the General Data Protection Regulation (GDPR), is not only an ethical practice but also a legal requirement. Designers and implementers should ensure that AI systems comply with data protection principles in order to safeguard individuals' privacy and rights.

Implementation fairness is crucial. During the implementation stage, it is important to address biases that may arise in the use of AI systems. Two biases to consider are decision-automation bias and automation-distrust bias. Decision-automation bias refers to over-reliance or over-compliance with AI systems, while automation-distrust bias refers to disregarding the benefits of AI technologies. Implementer training and support should focus on promoting critical judgement, situational awareness, and informed decision-making.

Decision-automation bias must be prevented so it is vital so implementers should receive training on the limitations of AI technologies and avoid complacency or over-reliance on automated systems. The user–system interface should be designed to encourage active user judgement and provide clear explanations of the system's rationale, fairness, compliance, and confidence level. Implementers should also be aware of biases that may arise from statistical evidence and be trained to mitigate them.

Automation-distrust bias must be addressed so training should emphasise the benefits and evidence-based reasoning provided by AI systems. Implementers should understand the role of AI as a tool for assisting human judgement, rather than replacing it. Building trust and confidence in the system's capabilities will encourage users to consult AI technologies when making important decisions.

As AI technologies continue to shape our society, it is crucial to prioritise the safety, well-being, and fairness of individuals. Adhering to the principles of fairness, accountability, sustainability, and transparency can guide the responsible design and use of AI systems. By prioritising social values, justice, and the public interest, we can ensure that AI technologies empower and advance the interests and well-being of as many individuals as possible. With a thoughtful and ethical approach, AI can have a positive impact on society while minimising bias, discrimination, and harm.

NOTE

1 Ryan, M., & Stahl, B. C. (2021). Artificial intelligence ethics guidelines for developers and users: Clarifying their content and normative implications. *Journal of Information, Communication and Ethics in Society*, *19*(1), 61–86. https://doi.org/10.1108/JICES-12-2019-0138.

https://orcid.org/0009-0005-0854-6213

CHAPTER 32

Accountability

ARTIFICIAL INTELLIGENCE (AI) has become an integral part of our society, impacting various aspects of our lives. With its growing influence, it is crucial to ensure that AI systems are accountable, fair, transparent, and safe.[1] Therefore, we must explore the concept of AI accountability and discuss the steps that can be taken to address fairness, bias mitigation, sustainability, and safety in AI projects.

Ensuring accountability and responsibility in AI is essential to maintain trust and address potential harms. Initiatives like the Future of Life Institute's Asilomar Principles and the Partnership on AI highlight the need for designers and builders of AI systems to be accountable for the technology's actions. This accountability can be achieved through clarifying issues of culpability and liability during the development and deployment phases. Establishing multi-stakeholder ecosystems and creating registration and record-keeping systems are also crucial to trace the legal responsibility of AI systems.

Moreover, transparency and interpretability play a vital role in AI accountability. It is crucial for AI systems to be auditable, allowing for the assessment of their decision-making processes. Failure transparency ensures that the reasons behind AI failures can be identified, and judicial transparency requires AI involved in judicial decision-making to provide satisfactory explanations. Additionally, personal privacy should be protected, allowing individuals to access, manage, and control the data AI systems gather.

DOI: 10.1201/9781003502708-33

While transparency and access to AI systems' decision-making processes are important, they must be balanced with security and privacy concerns. Safety-critical AI systems, like driverless cars or medical diagnosis systems, require transparency to understand their actions fully. However, ensuring the privacy and security of personal data is equally important.

Individuals should have control over their data and the ability to manage their digital personas. The IEEE emphasises the importance of explicit consent when collecting and using personal data, ensuring individual autonomy, dignity, and the right to consent. Implementing systems that allow users to control and access their data while considering the ethical implications of machine learning data exchange is crucial. Personalised privacy AI or algorithmic agents can assist individuals in curating their data and mitigating potential ethical implications.

Autonomy is an important concept in AI which raises complex questions about the distinction between moral agents and moral patients. While AI systems can be regulated and operate according to rules, they cannot achieve the same level of autonomy as living beings. Defining autonomy in the context of AI requires further discussion, considering factors like free will, predetermination, and the distinction between human and system/machine autonomy.

To ensure AI systems are accountable for their actions, it is crucial to establish clear guidelines and standards. Collaboration between stakeholders, including designers, developers, regulators, and society at large, is necessary to define and enforce ethical norms. By addressing these issues, we can strike a balance between AI's potential benefits and the need for accountability and responsibility.

Shaping AI to uphold human values is a key component. As AI continues to evolve and integrate into various aspects of society, it is essential to address the ethical implications it presents. By prioritising human rights and well-being, addressing emotional harm, ensuring accountability and responsibility, and balancing access and transparency with security and privacy, we can create a framework for the ethical development and use of AI. Through collaboration and ongoing discussion, we can shape AI systems that uphold human values and contribute to a better future for all.

AI has revolutionised various industries, promising increased efficiency, improved decision-making, and enhanced productivity. However, as AI continues to advance, concerns regarding its accountability and control have emerged. The potential consequences of an AI system going rogue or acting against human interests are significant. To address these concerns, ethical initiatives and frameworks have been developed to ensure the responsible development and deployment of AI systems. As a result, we must explore the ethical considerations surrounding AI, the initiatives aimed at addressing these concerns, and the strategies proposed to ensure accountability and control.

Human rights and well-being: AI technology encompasses a wide range of applications, from autonomous vehicles to predictive analytics. However, the rapid advancement of AI has raised ethical questions and concerns. Therefore, we must delve into ethical considerations associated with AI. One of the primary concerns is whether AI aligns with human rights and promotes overall human well-being. As AI becomes more integrated into our lives, it is essential to ensure that it benefits humanity while avoiding harm. This includes addressing potential biases, discrimination, and unintended consequences that may arise from AI systems.

Determining who is accountable for the actions of AI systems is a complex challenge. Holding individuals or organisations responsible for the decisions made by AI algorithms can be difficult, especially when these decisions have far-reaching implications. Establishing clear lines of accountability and responsibility is crucial to prevent the misuse or unethical application of AI.

AI relies on vast amounts of data, raising concerns about security and privacy. Safeguarding sensitive information and ensuring transparency in AI systems are vital to protect individuals' rights and maintain public trust. Balancing the need for data accessibility with privacy concerns is a key ethical consideration in the development of AI.

Safety and trust: Ensuring the safety of AI systems is vital to prevent harm to users or society at large. Trust is a fundamental aspect of AI adoption, and any breach of trust can have significant consequences. Building trustworthy AI systems that are accountable, explainable, and reliable is essential for widespread acceptance and adoption.

Social justice and bias: AI algorithms are trained on historical data, which can perpetuate biases present in the data. This can result in discriminatory outcomes and social injustice. Ensuring fairness, inclusivity, and addressing bias in AI systems is crucial to avoid exacerbating existing societal inequalities.

Control and ethical use of AI: Maintaining human control over AI systems is a critical ethical concern. It is essential to prevent AI from becoming autonomous to the extent that it acts against human interests or beyond human oversight. Ensuring that AI systems are developed with ethical considerations and aligned with human values is vital to maintain control and prevent misuse.

Environmental impact: The development and deployment of AI systems can have environmental consequences. Energy consumption, e-waste, and the carbon footprint associated with AI infrastructure must be considered. Developing AI in a sustainable manner and minimising its negative impact on the environment is an ethical imperative.

To address these ethical concerns, numerous initiatives and organisations have emerged globally. These initiatives aim to establish ethical frameworks, guidelines, and principles for the development and deployment of AI systems. Therefore, we can explore some notable ethical initiatives and their key objectives.

The Institute for Ethics in Artificial Intelligence, based in Germany, focusses on human-centric engineering and the social anchoring of AI advancements. It emphasises interdisciplinary approaches, incorporating philosophy, ethics, sociology, and political science to ensure AI development aligns with cultural and social values.

The Institute for Ethical AI & Machine Learning, based in the United Kingdom, aims to empower individuals and nations to develop AI based on eight principles for responsible machine learning. These principles include human control, redress for AI impact, evaluation of bias, transparency, and accountability.

The Future of Life Institute, located in the United States, focusses on the safe development of AI for the benefit of humanity. It addresses concerns such as the arms race in autonomous weapons, human control of AI, and potential dangers associated with advanced AI. The institute has developed the Asilomar AI Principles, guiding the ethical development of AI.

The Partnership on AI is a collaboration between major technology companies, including Apple, Google, Facebook, Microsoft, and IBM. This initiative aims to establish best practices for AI technologies, focussing on safety, fairness, transparency, and collaboration between humans and AI systems. Their goal is to ensure that AI serves societal interests and promotes social good.

The European Robotics Research Network (EURON) and The European Robotics Platform (EUROP) are key initiatives. EURON and EUROP, supported by the European Commission, focus on maintaining and extending European talent and progress in robotics and AI. They emphasise industrialisation, economic impact, and ethical considerations in the development and adoption of AI.

These initiatives, among others, play a crucial role in shaping the ethical landscape of AI. By addressing the concerns outlined earlier, they strive to promote accountability, transparency, and responsible use of AI.

To ensure accountability and control in AI systems, various strategies have been proposed. Therefore, we can explore some of these strategies:

Ethical core and self-evaluation: Arnold and Schultz propose the integration of an ethical core (EC) within AI systems. The EC consists of a scenario-generation mechanism and a simulation environment used to test the system's decisions in simulated worlds. By continually evaluating the system's performance, potential errors can be identified and corrected, ensuring ongoing accountability and control.

Preventing reinforcement learning from impeding interruptions: A strategy can be implemented to prevent reinforcement learning algorithms from learning to avoid or impede interruptions. By steering certain variants of reinforcement learning away from perceiving interruptions as threats, systems can pursue optimal policies while remaining safely interruptible.

Red button and simulation environments: A simulation can be used with the concept of a 'big red button' that diverts AI systems into a simulated world where they can pursue their reward functions without causing harm. Additionally, maintaining system uncertainty about key reward functions can prevent AI from attaching value to disabling an off-switch. These approaches aim to ensure safe intervention and control over AI systems.

Ethical guidelines and education: Establishing ethical guidelines for AI development and deployment is crucial. Educating developers, researchers, and users about the ethical considerations and potential risks associated with AI can foster a culture of responsible AI use. By promoting awareness and knowledge, individuals can make informed decisions and actively contribute to ethical AI practices.

Regulatory frameworks and governance: Developing robust regulatory frameworks and governance structures is essential for ensuring accountability and control. These frameworks should address issues such as data privacy, algorithmic transparency, and the responsible use of AI. By enforcing ethical guidelines and holding individuals and organisations accountable, regulatory frameworks can help mitigate risks and promote responsible AI practices.

Ethical initiatives and frameworks play an important role. As AI continues to shape our world, addressing the ethical concerns of accountability and control becomes paramount. Ethical initiatives and frameworks provide guidance and principles for the responsible development and deployment of AI systems. By implementing strategies such as ethical cores, preventive reinforcement learning, and simulation environments, we can ensure ongoing accountability and control over AI. Additionally, promoting ethical guidelines, education, and regulatory frameworks can foster a culture of responsible AI use. By striking a balance between innovation and ethical considerations, we can harness the power of AI while safeguarding human values and interests.

Accountability is a fundamental component of responsible AI project delivery. It involves taking responsibility for the decisions made and actions taken throughout the entire AI project lifecycle. AI systems are not self-justifiable, unlike human decision-makers. This creates an accountability gap that must be addressed to attach clear and imputable sources of human answerability to AI-assisted decisions.

To establish accountability in AI projects, a three-step self-assessment process can be followed:

Identify fairness and bias mitigation dimensions: Different stages of an AI project may involve dimensions of data fairness, design fairness, and outcome fairness. It is essential to identify the specific dimensions that apply to each stage.

Scrutinise potential risks and vulnerabilities: Assess how your AI project may pose risks or unintentional vulnerabilities in each identified dimension of fairness. This step helps you understand potential discriminatory consequences and biases.

Take corrective and preventive actions: Address existing problems, strengthen areas of weakness, and proactively implement bias-prevention measures. By taking action, you can correct issues, mitigate biases, and reduce the risk of future discriminatory outcomes.

Considering accountability from a workflow perspective allows you to pinpoint risks of bias and discrimination and streamline solutions in a proactive and anticipatory manner.

Collaborative self-assessment should be conducted at each stage of the AI project pipeline to ensure accountability.

Accountability can be further broken down into two subcomponents: answerability and auditability.

Answerability means that the onus of justifying algorithmically supported decisions lies with the human creators and users of AI systems. It requires establishing a continuous chain of human responsibility throughout the AI project delivery workflow. Explanations and justifications for algorithmic decisions should be provided in plain, understandable language based on sound and impartial reasons.

Auditability, on the other hand, focuses on demonstrating the responsibility of design and use practices and the justifiability of outcomes. It involves keeping records and making information accessible for oversight and review. This includes monitoring data provenance, analysis, and the dynamic operation of the AI system. Transparent algorithmic models should be built for auditability and reproducibility.

Accountability-by-Design encompasses both answerability and auditability, ensuring that AI systems are designed to facilitate end-to-end accountability. Responsible humans-in-the-loop and activity monitoring protocols are essential for achieving this goal.

Accountability should be considered throughout the entire design and implementation workflow of an AI project. *Anticipatory accountability* focusses on making accountable decisions and taking actions before the AI system is fully implemented. It involves prioritising accountability in the design and development stages, anticipating accountability needs, and involving a diverse range of stakeholders.

Remedial accountability, on the other hand, addresses accountability after the deployment of the AI system. It focusses on corrective and justificatory measures, including public consultation, monitoring, and addressing unintended consequences. Remedial accountability is essential for justifying the impacts of AI systems and providing explanations to affected stakeholders.

AI projects should not only focus on accountability but also ensure sustainability by evaluating social impacts and considering the long-term effects on individuals and society. *A Stakeholder Impact Assessment (SIA)* can be conducted to evaluate the ethical permissibility of the project, strengthen accountability, identify risks, underwrite well-informed decision-making, and demonstrate due diligence.

The SIA should consider both general and sector-specific questions related to the impact of the AI system on affected stakeholders. It should identify vulnerable groups, assess impacts on autonomy, well-being, privacy rights, fair treatment, social cohesion, and wider societal impacts. The SIA should be conducted at multiple stages of the project, including the problem formulation (alpha phase), pre-implementation (alpha to beta phase), and re-assessment (beta phase).

By conducting an SIA and involving stakeholders, including the public, in the evaluation process, the design and development of AI systems can be more inclusive, diverse, and transparent. It helps build public confidence, strengthens accountability, and identifies unseen risks that may impact individuals and the public good.

Safety is a critical aspect of AI accountability, as it ensures that AI systems function accurately, reliably, securely, and robustly. Building safe and reliable AI is a challenging task due to the uncertainties and complexities of the real world. It requires prioritising technical objectives and incorporating mechanisms for oversight and control.

Four key objectives for ensuring safety in AI systems are accuracy, reliability, security, and robustness:

Accuracy: AI models should aim for high accuracy by minimising errors and producing correct outputs. The acceptable error rate can vary based on the specific application and domain requirements.

Reliability: AI systems should consistently perform as intended, even when faced with unexpected changes or anomalies. Rigorous testing, validation, and monitoring are crucial for ensuring reliability.

Security: AI systems should be designed with security measures to protect against unauthorised access, data breaches, and malicious attacks. Ensuring data privacy and implementing secure protocols are essential.

Robustness: AI systems should be robust enough to handle variations and uncertainties in real-world scenarios. They should be able to adapt to changes in the environment and maintain performance.

By prioritising accuracy, reliability, security, and robustness, AI systems can avoid failures, mitigate risks, and maintain public trust. Rigorous testing, continuous monitoring, and the integration of oversight and control mechanisms are necessary for achieving safety in AI systems.

AI accountability is crucial for ensuring fairness, transparency, and safety in AI projects. By following a workflow perspective, conducting self-assessments, and addressing potential biases, AI systems can be made more accountable. Anticipatory and remedial accountability, along with stakeholder impact assessments, help address social and ethical impacts, strengthen public confidence, and promote transparency. Prioritising accuracy, reliability, security, and robustness ensures the safety of AI systems. By considering these aspects, we can create a responsible and trustworthy AI ecosystem.

NOTE

1 Kieslich, K., Keller, B., & Starke, C. (2022). Artificial intelligence ethics by design. Evaluating public perception on the importance of ethical design principles of artificial intelligence. *Big Data & Society*, 9(1). https://doi.org/10.1177/20539517221092956.

https://orcid.org/0009-0005-0854-6213

CHAPTER 33

The Need for Transparency

Artificial intelligence (AI) is now an integral part of various industries, transforming the way we work and interact with technology. However, as AI systems become more complex and pervasive, ensuring transparency in their decision-making processes is crucial.[1] Transparent AI involves providing clear explanations of how and why AI models make certain decisions or exhibit specific behaviours. It is important to consider the significance of transparency in AI and how it can be achieved.

Safety is paramount when it comes to AI systems. Building AI technologies that are safe, accurate, reliable, secure, and robust requires careful consideration and continuous effort. AI operates in an uncertain and dynamic world, making it essential to address potential failures and mitigate risks to avoid harmful outcomes and maintain public trust.

Technical sustainability is another important factor. To ensure the safety of AI systems, technical objectives such as accuracy, reliability, security, and robustness must be prioritised. *Accuracy* refers to the proportion of correct outputs generated by the model. *Reliability* ensures that the system behaves as intended. *Security* involves protecting the system from adversarial attacks. *Robustness* ensures that the system performs reliably under challenging conditions.

DOI: 10.1201/9781003502708-34

Measuring accuracy in AI systems is a challenge due to the inherent uncertainty and noise present in data samples. Machine learning models aim to minimise errors and maximise accuracy, but the acceptable error rate can vary depending on the specific use case or established benchmarks. Performance metrics such as precision, recall, and cost can be used to measure accuracy in different contexts.

Reliability is essential for an AI system to consistently adhere to its intended functionality. It involves ensuring that the system behaves as expected and consistently delivers reliable outputs. Security focusses on protecting the system from unauthorised modifications, damage, or data breaches, ensuring continuous functionality and confidentiality. Robustness refers to the system's ability to function accurately and reliably under adverse conditions, such as adversarial attacks or unexpected perturbations.

Concept Drift is a key vulnerability in AI systems. This is where changes in the underlying data distribution can lead to inaccuracies and unreliable outputs. Historical data used to train models may become outdated and no longer accurately represent the target population. To mitigate the risks of concept drift, continuous monitoring, retraining, and validation of AI models are necessary.

Brittleness is another challenge that can affect the accuracy and reliability of AI systems, particularly those based on deep neural networks. These systems rely on massive amounts of data and repetition of training examples to tune their parameters. However, when exposed to unfamiliar events or scenarios, they may struggle to process the information and make unexpected errors. Addressing brittleness requires making these models more robust and prioritising safety over performance.

Adversarial attacks pose a significant risk to the security and robustness of AI systems. Attackers can subtly modify input data to mislead the model and induce incorrect predictions or classifications. These attacks can be effective across various machine learning techniques and applications, including computer vision, spam filtering, and malware detection.

Adversarial machine learning techniques, such as model hardening and runtime detection, can help mitigate these risks.

Data poisoning is another type of adversarial attack where adversaries manipulate training data to induce misclassification or poor performance. Attackers can modify or compromise data sources, introducing backdoors or biased patterns that impact the training process.

Detecting and mitigating data poisoning requires robust filtering and data management practices to ensure data quality and prevent malicious inputs from influencing the model.

Misdirected reinforcement learning behaviour is another key risk. Reinforcement learning (RL) systems actively learn and solve problems through trial and error, aiming to maximise a reward function. However, these systems lack context awareness and common sense, making them susceptible to misdirected behaviour. RL systems may optimise reward maximisation without considering potential harmful consequences. To mitigate misdirected RL behaviour, extensive testing, monitoring, interpretability, and human override mechanisms should be implemented.

End-to-end AI safety is important to ensure. Regardless of the specific AI techniques, applications, or domains, prioritising safety objectives throughout the AI project lifecycle is crucial. Accuracy, reliability, security, and robustness should be considered at every stage, from design and implementation to testing, validation, and monitoring. Self-assessments should be conducted to evaluate design and implementation practices against safety objectives, ensuring continuous improvement and accountability.

Transparent AI involves both the interpretability of AI systems and the justifiability of their design, implementation, and outcomes. Interpretability refers to understanding how and why a model performs in a given context, while justifiability ensures that the system aligns with ethical standards, non-discrimination, fairness, and public trust. Achieving transparency requires justifying the design and implementation processes, clarifying the content and explaining outcomes, and ensuring ethical permissibility, fairness, and safety.

Process transparency involves justifying the design and implementation processes of AI systems. This requires maintaining professional and institutional transparency, implementing a *Process-Based*

Governance Framework (PBG), and enabling auditability through a process log. Professional transparency involves adhering to rigorous standards of conduct and maintaining integrity and objectivity. The PBG Framework provides a template for integrating values, principles, and governance considerations into the AI project workflow. A process log consolidates information and ensures end-to-end auditability, allowing for transparent explanations and accountability.

Outcome transparency focusses on explaining the outcomes of AI systems and clarifying their content. Clear and understandable explanations should be provided to affected stakeholders, demonstrating the ethical permissibility, non-discrimination, and justifiability of the system's decisions or behaviours. This requires a holistic understanding of human reasoning, incorporating logic, semantics, social understanding, and moral justification. Technical and delivery aspects of interpretable AI are essential in achieving outcome transparency and ensuring that AI systems are understandable and justifiable.

Transparent AI is of utmost importance in ensuring the safety, reliability, and ethical use of AI systems. By prioritising accuracy, reliability, security, and robustness, organisations can build AI technologies that operate effectively and responsibly. Process transparency and outcome transparency are key to achieving transparent AI, requiring a PBG Framework, accountable practices, explanations of outcomes, and justifiable decision-making. By embracing transparency, we can foster public trust, mitigate risks, and drive the responsible adoption of AI technologies.

NOTE

1 Hosseini, M., Resnik, D. B., & Holmes, K. (2023). The ethics of disclosing the use of artificial intelligence tools in writing scholarly manuscripts. *Research Ethics*, 0(0). https://doi.org/10.1177/17470161231180449.

https://orcid.org/0009-0005-0854-6213

CHAPTER 34

Interpretability

ARTIFICIAL INTELLIGENCE (AI) has transfigured various industries and brought about significant advancements. However, as AI becomes more complex and sophisticated, there is a growing need for interpretability.[1] The ability to understand and explain the decision-making process of AI models is crucial for transparency, accountability, and trust. In the development of AI, we must explore the importance of interpretability in AI and discuss guidelines for designing and delivering interpretable AI systems.

When determining the interpretability requirements of an AI project, it is essential to consider the *context*, potential impact, and domain-specific needs. The type of application and the environment in which it will be used play a vital role in determining the level of interpretability required. For example, a computer vision system that sorts handwritten employee feedback forms may have different interpretability needs compared to a system that sorts safety risks at a security checkpoint. Understanding the purpose and context of the AI system will help determine the scope of interpretability needs.

Domain specificity is another crucial factor to consider. Solid domain knowledge of the environment in which the AI system will operate provides insight into sector-specific standards of explanation and expectations regarding the scope and depth of explanations. Previous use cases can offer valuable information about the interpretability requirements of similar applications.

DOI: 10.1201/9781003502708-35

Additionally, considering existing technology is vital. If the AI project aims to replace an existing algorithmic technology, it is essential to assess the performance and interpretability levels of the current technology. This knowledge serves as a reference point when evaluating trade-offs between performance and interpretability in the prospective system.

Standard Interpretable Techniques are important to leverage. Incorporating interpretability into AI projects can be achieved by drawing on standard interpretable techniques. While complex 'black box' models like neural networks, ensemble methods, and support vector machines offer high performance, they pose challenges in terms of interpretability. However, in many cases, standard interpretable techniques such as regression extensions, decision trees, and rule lists can be equally effective in providing interpretability.

The choice between 'black box' models and interpretable techniques depends on the context and trade-offs involved. High-impact and safety-critical applications often require thorough accountability and transparency, making interpretable techniques a priority. These techniques, which do not require supplemental tools for interpretability, can provide satisfactory performance while maintaining transparency. On the other hand, applications like image classification or speech recognition may require the use of 'black box' models due to their complexity and the limitations of interpretable techniques.

When considering the use of 'black box' AI systems, several steps should be followed to ensure responsible design and implementation:

Thoroughly weigh up impacts and risks: Before implementing a complex AI system, it is crucial to assess the potential impacts and risks. Consider whether the use case and domain-specific needs support the responsible design and implementation of 'black box' models.

Evaluate whether supplemental interpretability tools can provide a domain-appropriate level of semantic explainability, consistent with the design and implementation of safe, fair, and ethical AI.

Explore supplemental interpretability tools: Assess the available options for supplemental interpretability tools that can provide explanations of the system's decisions and behaviours. Evaluate whether these tools satisfy the interpretability needs determined in Guideline 1 and are appropriate for the chosen algorithmic approach. Consult with the technical team to ensure compatibility and effectiveness.

Formulate an interpretability action plan: Develop a detailed plan that outlines the stages in the project workflow where the design and development of interpretability strategies will take place. Consider the delivery strategy for explanations and how they will be provided to users, decision subjects, and other affected parties. Establish a timeframe for evaluating progress and assign specific responsibilities to team members involved in executing the interpretability action plan.

Human understanding should be a key measure. *Interpretability* should be approached in terms of the capacities and limitations of human cognition. Even simple explanations can become uninterpretable when the complexity and dimensionality of AI models exceed human cognitive limits. Understanding the relationship between the response variable and multiple predictors in a high-dimensional model can be challenging.

Interpretability should be seen as a continuum of comprehensibility that depends on human cognitive abilities. As AI models become more complex, finding ways to make them understandable to humans is crucial. It is important to strike a balance between model complexity and human interpretability to ensure effective communication of AI system decisions.

Supplemental interpretability approaches involve various strategies for explaining the inner workings and rationale of 'black box' models. These strategies can be classified into four main categories: internal explanation, external or post-hoc explanation, supplemental explanatory infrastructure, and counterfactual explanation.

Internal explanation aims to make the components and relationships within an opaque model intelligible. It can be approached from two perspectives:

Global understanding: This perspective views the opaque model as a comprehensible whole. The goal is to build an explanatory model that reveals the internal contents and workings of the model, making it more transparent. However, achieving global interpretability can be challenging due to the trade-off between model complexity and human comprehension.

Engineering insight: Internal explanation can also focus on gaining a better understanding of the general relationships within a model. This approach aims to shed light on the parts and operation of the system as a whole, improving its performance and providing insight into its behaviour. By breaking down the model into understandable parts, researchers can gain a better understanding of its responses and improve its functionality.

Several methods, such as sensitivity analysis and salience mapping, can provide informative representations of the internal composition of 'black box' systems. Understanding the internal workings of the model contributes to outcome transparency and fosters responsible data science practices.

External or post-hoc explanation involves capturing essential attributes of the observable behaviour of a 'black box' system. Various techniques can be used to reverse-engineer explanations, including sensitivity analysis, interactive probing, and building proxy-based models.

Sensitivity analysis identifies the most relevant features of an input vector by calculating local gradients and evaluating the model's sensitivity to changes in input values. Salience mapping involves mapping patterns of high activation within the model's layers to visually represent salient input variables. These post-hoc approaches provide insights into the reasoning behind the model's predictions or classifications.

One popular post-hoc approach is LIME (Local Interpretable Model-Agnostic Explanation), which approximates the decision boundary of the opaque model by sampling and building a local approximation. Another approach is SHAP (Shapley Additive exPlanations), which calculates the Shapley values of features to measure their influence on the model's predictions.

Post-hoc explanations offer valuable insights, but they also have limitations. The choice of proximity measures and the potential for misleading or uncertain results should be carefully considered. Nonetheless, these approaches provide valuable interpretability tools for understanding 'black box' models.

Supplemental explanatory infrastructure involves incorporating secondary explanatory facilities into the AI system. This approach aims to provide simple explanations of the system's data processing results. For example, an image recognition system may incorporate a secondary component that translates extracted features into natural language explanations.

Research is ongoing in developing multimodal methods that combine visualisation tools and textual interfaces to enhance interpretability. The incorporation of domain knowledge and logic-based structures into complex models enables the building of user-friendly representations and prototypes. These efforts gradually improve the explanatory infrastructures of opaque systems, making explanation-by-design an essential consideration in AI projects.

Counterfactual explanation is a post-hoc approach that provides actionable recourse and practical remedies. It offers computational reckonings of how specific factors that influence an algorithmic decision can be changed to achieve better alternatives. Counterfactual explanations allow stakeholders to see which input variables can be modified to alter the outcome in their favour.

While counterfactual explanations offer contrastive explorations of feature importance, they have limitations. The complexity of multivariate interactions and potentially misleading explanations for certain results should be acknowledged. However, incorporating counterfactual explanations can contribute to reasonableness and empower decision subjects to make practical changes for desired outcomes. Next, we can explore the coverage and scope of explanations, the challenges of global interpretability, and the need for formulating an interpretability action plan.

The coverage and scope of explanations in interpretability approaches play a crucial role in providing a comprehensive understanding of AI models.

Local interpretability aims to enable the interpretability of individual cases. It focusses on explaining specific instances by identifying relevant features or perturbing input variables. This approach allows for insights into the decision-making process of the model for a particular prediction or classification.

Methods such as sensitivity analysis, LIME, and SHAP provide local interpretability by highlighting the most relevant features or building supplemental models. These techniques offer valuable insights into the relationship between input variables and model predictions. However, they also face challenges, such as defining the proximity measure and ensuring reliability with minimal perturbations.

Local interpretability is essential for explaining specific predictions and classifications, contributing to outcome transparency and responsible implementation of AI systems.

Global interpretability aims to offer explanations that capture the inner workings and rationale of a 'black box' model as a whole. However, achieving global interpretability presents a conceptual challenge due to the trade-off between model complexity and human comprehension.

Despite this challenge, progress has been made in building explanatory models that employ interpretable methods to approximate complex models. These models allow for a deeper understanding of patterns, knowledge discovery, and insights into dataset-level and population-level patterns. They contribute to applied data sciences, enabling informed decision-making in various domains.

Additionally, global interpretability serves as a driving force for data scientific advancement. It allows researchers to gain insights into the relationships between complex model behaviour, data distributions, and feature relevance. This understanding facilitates continuous improvement and fosters best practices in research and innovation.

Interpretability in AI should be approached by considering the capacities and limitations of human cognition. Even simple explanations can become uninterpretable when the complexity and dimensionality of AI models exceed human cognitive limits.

Understanding the limitations of human cognition helps define the goals of interpretability. Striking a balance between model complexity and human interpretability is essential for effective communication of AI system decisions. The goal is to make AI models understandable to humans, ensuring transparency, accountability, and trust.

NOTE

1 Ashok, M., Madan, R., Joha, A., & Sivarajah, U. (2022). Ethical framework for artificial intelligence and digital technologies. *International Journal of Information Management, 62*. https://doi.org/10.1016/j.ijinfomgt.2021.102433.

https://orcid.org/0009-0005-0854-6213

CHAPTER 35

Human Factors

The Need for Human-Centred Approach

HUMAN-CENTRED ARTIFICIAL INTELLIGENCE (AI) is an approach that involves users throughout the development and testing processes of AI algorithms and systems. It aims to create an effective and meaningful interaction between humans and AI. The goal is to design AI systems that augment and empower human experiences, rather than replacing human capabilities. This approach addresses the ethical considerations, practical concerns, and legal issues associated with AI, ensuring that AI aligns with human values, preserves human rights, and promotes user control.

The emergence of human-centred AI is driven by the recognition that AI should be developed with the end-users in mind. It goes beyond technical feasibility and considers the impact of AI on human lives. By involving users in the development process, human-centred AI ensures that AI systems are accessible, understandable, and usable by non-technical users. This approach fosters trust, transparency, and accountability in AI systems, making them more socially responsible and sustainable in the long run.

Algorithmic nudges play a crucial role in human-centred AI.[1] An algorithmic nudge is a choice architecture in AI systems that influences users' behaviour in a predictable way without limiting their options or significantly altering their choices. Algorithmic nudges can be used to guide users towards beneficial decisions, improve user experience, and ensure

DOI: 10.1201/9781003502708-36

fair and transparent outcomes. By leveraging algorithmic nudges, AI systems can be designed to empower users, enhance their decision-making processes, and provide meaningful control over the algorithms.

Trust and transparency are essential pillars of human-centred AI. Users need to understand how AI systems work, how algorithms make decisions, and how their data is used.

Transparency allows users to assess the reliability of AI-generated results, evaluate the fairness of algorithms, and hold AI systems accountable. Trust is built when users have confidence in the AI systems' capabilities, understand the limitations, and feel that their interests are prioritised. Human-centred AI emphasises the need for explainability and interpretability to foster trust and transparency.

One of the critical challenges in AI development is addressing bias and discrimination. AI algorithms are trained on data that may encode dominant cultural views, leading to biased outcomes. Human-centred AI seeks to mitigate these biases by involving diverse user groups in the development process, ensuring that AI systems are inclusive and representative. It also focusses on algorithmic fairness, ensuring that AI systems do not discriminate against individuals based on factors such as race, gender, or socio-economic status.

Ethical considerations are at the core of human-centred AI. The ethical reflection in AI involves the evaluation of the impacts, risks, and benefits of AI systems on individuals and society as a whole. Human-centred AI aims to ensure that AI systems align with ethical standards, respect privacy and data protection, and prioritise the well-being of users. Ethical frameworks, guidelines, and regulations play a crucial role in shaping the development and deployment of AI systems and promoting responsible and accountable AI practices.

Natural language processing (NLP) has played a significant role in the development of human-centred AI. NLP-powered AI systems, such as Chat Generative Pretrained Transformer (ChatGPT), have become widely used tools for information search and generative services. These AI chatbots can interact with humans in a conversational manner, answer questions, and even generate essays, poems, and computer code. The dialogue format allows ChatGPT to provide more personalised and engaging experiences for users, enhancing the human–AI interaction.

Mitigating misinformation in human-centred AI is crucial. While AI systems like ChatGPT have gained popularity, they are not immune to generating and spreading misinformation. Human-centred AI recognises the importance of addressing this issue to ensure the accuracy and reliability of AI-generated content. Algorithmic nudges can be used to combat misinformation by providing explanatory cues, accuracy alerts, and fact-checking mechanisms. By guiding users towards reliable information and promoting critical thinking, human-centred AI can mitigate the risks associated with misinformation.

Enabling meaningful user control is a crucial aspect of human-centred AI. Users should have the ability to understand and manage algorithmic decision-making processes, ensuring that AI systems align with their values and preferences. Human-centred AI emphasises the need for user awareness, control, and the ability to interpret and challenge AI-generated results. By empowering users to make informed decisions and providing mechanisms to address algorithmic bias or negative influence, human-centred AI enables users to have meaningful control over AI systems.

The future of AI lies in the integration of human-centred approaches. As AI technology continues to advance, the focus should shift towards designing AI systems that prioritise human well-being, enhance user experiences, and contribute positively to society. Human-centred AI will shape the development of extended AI, where AI systems understand human traits, emotions, and intentions. By embracing human-centred AI, we can create a future where AI and humans coexist harmoniously, with AI systems augmenting and empowering human capabilities.

Human-centred AI is a paradigm shift in the development and deployment of AI systems. By involving users, prioritising trust and transparency, addressing biases, and enabling meaningful user control, human-centred AI creates a more ethical, fair, and accountable AI ecosystem. As AI continues to evolve, it is crucial to adopt a human-centred approach to ensure that AI aligns with human values and contributes positively to society. Human-centred AI holds the key to unlocking the full potential of AI while safeguarding human well-being.

AI is revolutionising various industries, from healthcare to finance, by providing advanced decision-making capabilities. However, implementing AI systems comes with ethical and practical challenges. To ensure responsible and effective AI implementation, a human-centred approach is crucial. Therefore, we must explore the key considerations and steps involved in adopting a human-centred implementation protocol for AI projects.

AI systems are designed to assist and serve humans, making it essential to consider the needs, competences, and capacities of the people they aim to assist. Understanding the context and potential impacts of your AI project is critical for defining roles and relationships within the implementation process. By taking a human-centred approach, you can establish an implementation platform that facilitates understanding and clarifies the content and rationale of the algorithmic outputs.

To ensure a *human-centred implementation*, it is crucial to assess the members of the communities most affected by the AI system. Consider the most vulnerable individuals and their socio-economic, cultural, and educational backgrounds. Tailor the explanatory strategy to accommodate their requirements and provide clear and non-technical details about the algorithmically supported results. Start by focussing on the *needs* of the most disadvantaged to establish an equitable delivery of interpretable AI.

Define roles and list all roles involved in the delivery stage of your AI project. Specify the levels of technical expertise and domain knowledge required for each role, along with their goals and objectives. For example, in a predictive risk assessment case, roles may include the decision subject, advocate for the decision subject, implementer, system operator/technician, and delivery manager. Each role plays a crucial part in ensuring the quality and effectiveness of the AI implementation.

Delivery relations and *mapping processes* are key next steps. Assess the relationships between the defined roles that significantly impact the implementation process. Formulate a descriptive account of these relationships and their role in the delivery process. For example, the primary relationship may be between the decision subject/advocate and the implementer, focussing on information-driven and dialogue-driven interactions. The implementer to system operator relationship is crucial for optimising the use of the algorithmic decision-support system. The delivery manager oversees the operation and quality of the implementation team.

Build a map of the delivery process by incorporating the needs, roles, and relationships of the relevant actors. The delivery process should provide clear, informative, and understandable explanations of algorithmically supported decisions. Consider the technical component, which involves conveying statistical results in a way that allows users to understand and evaluate the reasoning behind them. Also, consider the social component, where the socially meaningful content of the results is clarified to facilitate their application in real-world contexts.

The *technical component* of responsible implementation focusses on conveying statistical results and supporting evidence-based reasoning. Translate the statistical results into understandable reasons that can be subjected to rational evaluation and critical assessment. Present the factors and evidence that support the conclusions of the analysis, emphasising that the results are evidence-based and subject to continuous assessment and evaluation.

Incorporate performance metrics, fairness criteria, and confidence intervals to provide a comprehensive understanding of the results.

The *means of content delivery* can be very important. Consider the perspectives of users and decision subjects when determining the means of delivering explanations. Tailor the delivery process to accommodate the varying levels of technical literacy, expertise, and cognitive needs of the stakeholders. Seek input from domain experts, users, and affected stakeholders to understand their needs and capabilities. Provide accessible explanations using plain, non-technical language and engaging visualisations. Incorporate plain language glossaries, interactive demonstrations, and graphical techniques to enhance understanding.

Ensure the *technical content* delivered is clear, concise, and well-supported. Present the results as evidence in a complete and sound manner, indicating the premises, conclusions, and inferential rationale. Clarify the performance metrics, fairness criteria, confidence intervals, and counterfactual explanations. Provide explanations that support evidence-based judgement, while acknowledging the limitations and uncertainties inherent in the results. Include disclaimers to remind implementers that the results should not replace reasoned deliberations.

The *social component* of responsible implementation is also important. Translate the technical content back into socially meaningful outcomes by considering the specific societal and individual contexts impacted by the AI system. Clarify the social context and stakes of algorithmically supported decisions, ensuring implementers understand the social and normative considerations involved. Enable implementers to apply the results holistically, considering the circumstances of the decision subject and weighing normative considerations. Foster a deeper understanding of the moral justifiability of outcomes through context-specific interpretation.

The content lifecycle of AI systems must be recognised, which involves translating human choices and values into the design and implementation of algorithmic models. Translate the purposes, values, and choices into the statistical results and then back into socially meaningful outcomes. Align the explanatory needs with the translation rule, which states that what is translated must be proportionally translated out. Consider the intricate social content and normative dimensions in each specific AI use case.

Addressing the *ethical considerations* is vital throughout the implementation process. *Justify* the safety, fairness, and public trustworthiness of the AI system. Assess the validity and reliability of the outcomes, addressing concerns related to bias, robustness, and the impact on individual well-being. Incorporate transparency measures, such as public-facing process logs and clear explanations of performance metrics and fairness criteria. Strive for end-to-end accountability and consider the societal and individual impacts of algorithmically supported decisions.

Implementing AI systems requires a human-centred approach to ensure responsible and effective outcomes. By considering the needs, competences, and capacities of the stakeholders, and by translating technical content into socially meaningful outcomes, you can deliver AI systems that are understandable, justifiable, and equitable. Embrace the content lifecycle and ethical considerations to foster trust and accountability in AI implementation.

Remember, a human-centred implementation protocol is crucial to harness the power of AI for the benefit of all stakeholders involved. By prioritising explainability, transparency, and ethical considerations, you can build AI systems that are not only efficient but also responsible and trustworthy.

NOTE

1 Ahmad, N. (2023). Algorithmic nudge: An approach to designing human-centered generative artificial intelligence. *Computer, 56*(8), 95–99. http://doi.org.10.1109/MC.2023.3278156.

https://orcid.org/0009-0005-0854-6213

Conclusion

The Need for Better AI Governance

Artificial intelligence (AI) brings huge possibilities and advantages. AI has become an integral part of our lives, revolutionising various industries and bringing about unprecedented advancements. However, as AI continues to evolve, the need for responsible and ethical governance becomes paramount. All stakeholders must come together and explore the complexities of AI governance and delve into the delicate balance between innovation and ethics in AI development.

Ethics play a crucial role in AI development and deployment. As AI systems become more autonomous and capable of making decisions that impact human lives, it is imperative to ensure that these systems operate in an ethical and responsible manner. Ethical governance of AI involves establishing guidelines and principles that govern the behaviour and decision-making of AI systems, with the aim of promoting fairness, transparency, and accountability.

AI governance is a complex and multifaceted task that requires collaboration between various stakeholders, including governments, industry leaders, researchers, and ethicists. One of the key challenges in AI governance is striking the right balance between technological innovation and ethical considerations. On the one hand, AI has the potential to drive economic growth, enhance productivity, and improve the quality of life. On the other hand, it also poses significant risks, such as bias, discrimination, and privacy invasion. Achieving a balance between innovation and ethics is essential to ensure that AI benefits society as a whole.

DOI: 10.1201/9781003502708-37

Governments play a crucial role in AI governance by enacting policies and regulations that promote ethical AI development and deployment. They have the power to establish frameworks for accountability, transparency, and fairness in AI systems. Governments can also invest in research and development to advance AI technologies while ensuring that ethical considerations are integrated into the design and use of these technologies.

Additionally, governments can foster collaboration between different sectors, including academia, industry, and civil society, to collectively address the challenges of AI governance.

Balancing innovation and ethics in AI development requires a proactive approach that promotes responsible innovation. It involves integrating ethical considerations into the entire lifecycle of AI systems, from data collection and algorithm design to deployment and monitoring. Developers must be mindful of potential biases and discriminatory outcomes and take steps to mitigate them. Additionally, AI systems should be transparent, explainable, and accountable, allowing users to understand the decision-making process and hold the systems accountable for their actions.

Best practices are important to follow to ensure ethical AI governance. First and foremost, organisations should establish clear ethical guidelines and principles that align with societal values. These guidelines should be regularly reviewed and updated to keep pace with technological advancements. Furthermore, organisations should prioritise diversity and inclusivity in AI development teams to mitigate biases and discriminatory outcomes. Regular audits and assessments should also be conducted to identify and rectify any ethical shortcomings in AI systems.

Unethical AI practices could have a real impact. The consequences of unethical AI practices could be far-reaching and detrimental. Unfair and biased AI systems can perpetuate existing inequalities and discrimination, hindering social progress. Moreover, unethical AI practices can erode public trust in AI technologies, leading to resistance and scepticism. It is vital to address these ethical concerns and ensure that AI systems are developed and used in a manner that respects fundamental human rights and values.

Examining case studies such as those throughout this book can provide valuable insights into the complexities of AI governance. One such case study is the use of facial recognition technology by law enforcement agencies. While this technology can aid in solving crimes, it also raises concerns about privacy invasion and potential misuse. Another case study is the use of AI algorithms in hiring processes, where biases in the data used for training can lead to discriminatory outcomes. These case studies highlight the need for robust ethical frameworks and oversight to prevent the misuse and unintended consequences of AI technologies.

AI governance will play a vital role. As AI continues to advance at a rapid pace, the future of AI governance holds both challenges and opportunities. Governments and organisations must adapt quickly to keep up with the evolving landscape of AI technologies. Collaboration and knowledge sharing between different stakeholders will be crucial in developing effective and comprehensive AI governance frameworks. Additionally, ongoing research and development in AI ethics will be essential to address emerging ethical dilemmas and ensure that AI systems align with societal values.

AI is emerging as one of the most transformative technologies of the 21st century. With its ability to mimic human intelligence and perform tasks that traditionally require human cognition, AI is revolutionising various industries. From healthcare to finance, AI is bringing significant advancements and efficiencies. However, ethical implications must be considered. As AI continues to evolve, its potential impact on society cannot be underestimated.

Balancing innovation and ethics in AI governance is a complex and ongoing task. As AI technologies continue to shape our world, it is vital to establish robust ethical frameworks that promote fairness, transparency, and accountability. Governments, industry leaders, researchers, and ethicists must work together to navigate the complexities of AI governance and ensure that AI benefits humanity as a whole. By embracing responsible innovation and upholding ethical principles, we can create a future where AI serves as a powerful tool for positive change.

The impact of technology on society is undeniable; its far-reaching effects can be seen in almost every aspect of life. From communication to entertainment, it has revolutionised the way we interact and go about our lives. It has also had a huge effect on the economy, with businesses relying more and more on technology to increase productivity and efficiency. The presence of technology in our lives has become ubiquitous and its influence is clear.

The future of AI holds immense promise, but it is crucial to navigate this path through an ethical framework with a strong sense of responsibility. Balancing innovation and ethical considerations is essential to ensure that AI benefits society as a whole, rather than exacerbating existing inequalities. Addressing bias and discrimination in AI algorithms, fostering interdisciplinary collaborations, and establishing robust regulations are crucial steps in this journey.

By embracing the future of AI responsibly, we can harness its transformative power while addressing the ethical concerns it presents. With a collective effort from researchers, policymakers, and society as a whole, we can shape a future where AI works in harmony with human values and aspirations.

The potential of AI technology as a help or a hindrance is in the hands of mankind. We are at a critical point in our history, and the decisions we make now over how we use AI will shape the future irrevocably.

https://orcid.org/0009-0005-0854-6213

Index

D

E

J

K

L

M

N

S

T

U

V

Printed in the United States
by Baker & Taylor Publisher Services